The British Navy, Economy and Society in the Seven Years War

The British Navy, Economy and Society in the Seven Years War

Christian Buchet

Translated by
Anita Higgie and Michael Duffy

THE BOYDELL PRESS

Originally published in French as *Marine, économie et societé. Un example d'interaction: l'avitaillement de la Royal Navy durant la guerre de sept ans*

First published 2013
The Boydell Press, Woodbridge

ISBN 978-1-84383-801-2

The Boydell Press is an imprint of Boydell & Brewer Ltd
PO Box 9, Woodbridge, Suffolk IP12 3DF, UK
and of Boydell & Brewer Inc.
668 Mount Hope Ave, Rochester, NY 14620-2731, USA
website: www.boydellandbrewer.com

A CIP catalogue record for this book is available from the British Library

Printed and bound in Great Britain by
CPI Group (UK) Ltd, Croydon, CR0 4YY

For Timothy and Aurélien

Contents

Foreword

Michael Duffy, University Fellow, formerly Director of the Centre for Maritime Historical Studies, University of Exeter

Following the original French language publication of this book in 1999,[1] it was hailed as a 'pioneering work' and it was admitted that 'no one had previously attempted a detailed analysis of the V[ictualling] B[oard]'s work or the tendering system by which it ensured that the navy was fed'.[2] Whereas historians had previously only considered naval victualling at best as a chapter in wider works,[3] Christian Buchet has afforded it an in-depth study covering its development over time and in particular its impact on Britain's most decisive imperial war of the eighteenth century – the Seven Years War. He has also explored its relations with the agricultural economy and with the contractors who made the agricultural market in the eighteenth century. He shows how it was possible to achieve reform in what has traditionally been regarded as the corrupt administrative system of the first half of the eighteenth century, and produce improvements that were both efficient and economic. Perhaps it took a French scholar who understood the workings of the French naval administration to see so clearly how these reforms provided a crucial difference which released forces to operate in large numbers over large distances and for long periods, whereas in the Seven Years War the French were unable to consider maintaining a large army in Canada because it was impossible to feed it there, while French naval expeditions to the Caribbean had barely time to arrive before lack of adequate provisions forced them to return home. Only

1 Christian Buchet, *Marine, économie et société. Un exemple d'interaction: l'avitaillement de la Royal Navy durant la guerre de sept ans* (Paris, Honoré Champion, 1999).

2 N. Rodger, *The Command of the Ocean: A naval history of Britain, 1649–1815* (London, Allen Lane, 2004), 398 n.50; J. Macdonald, 'A new myth of naval history? Confusing magnitude with significance in British naval victualling purchases, 1750–1815', *International Journal of Maritime History*, XXI (2009), 159–88, at 167.

3 E.g. D.A. Baugh, *British Naval Administration in the Age of Walpole* (Princeton, Princeton University Press, 1965), ch. 8; S.F. Gradish, *The Manning of the British Navy during the Seven Years War* (London, R.H.S., 1980), ch. 6; N.A.M. Rodger, *The Wooden World* (London, Collins, 1986), ch. 3; D. Crew, *Yellow Jack and the Worm. British Naval Administration in the West Indies 1739–1748* (Liverpool, Liverpool University Press, 1993), ch. 4.

the British had discovered the crucial secret of how to sustain large-scale war at a distance.

While we lack figures for its domestic competitors such as the East India Company, the Victualling Board was almost certainly the biggest single purchaser of agricultural produce in wartime, and though the London market where it obtained most of its needs was already the largest in Europe, the Victualling Board was not without influence on that market in its attention both to the quality and to the price of its purchases. Janet Macdonald has questioned the extent of that influence in contesting the claims of John Brewer and Nicholas Rodger, but at the very least, in wartime, the Navy became a significant player in the market economy both in London and around home outports as well as in the western Atlantic seaboard trade in foodstuffs. It has been argued that it did not increase the market as its seamen would have had to be fed anyway if they were not in the Navy, but it is likely that not so many would have been fed from London and nor would they have been fed with the amount and quality of food that the Navy required for its seamen. Macdonald concedes that the Victualling Board may have contributed to the shift towards wheat in the types of grain eaten and also to the rise in meat consumption.[4] Christian Buchet has been cautious in his claims, saying that 'the Navy both took advantage of the development of commercial capitalism and contributed to it', but he shows that the purchasing power of the Victualling Board enabled it to insist on high quality and also its readiness to employ competitive tender and even to resort to opening up new markets or refusing to buy if it thought prices were too high contributed to driving prices down even in wartime. In consequence it actually spent less in successfully victualling 70,000 men in 1758 during the Seven Years War than it did in victualling 40,000 men in 1710 during the War of the Spanish Succession. Moreover because the Board's credit was the soundest, and included adding interest when payment was delayed, its custom was much sought after, to the extent that over the course of the war there was a movement towards merchants forming associations to bid for the large contracts for which it invited tenders. The wartime victualling needs of the Navy did not create a national market but they enhanced it, as they also enhanced the Atlantic-wide trading system to sustain expeditions overseas. The most prominent beneficiaries of this were the large merchants with the resources and contacts to provide the enormous quantities required, and Christian Buchet has been able to shed light on exactly who these were.

4 Macdonald, 'A new myth of naval history?', 160–1, 164. Cf. J. Brewer, *Sinews of Power: War, money and the English State 1688–1783* (London, Unwin Hyman, 1989), 37, and N. Rodger, *The Command of the Ocean: A naval history of Britain, 1649–1815* (London, Allen Lane, 2004), 307.

Once he had shown the way, others followed with books and articles on the Victualling Board's operations in the wars against the French Revolution and Napoleon.[5] His work on the Seven Years War, however, still remains without competition as the definitive version and it is for this reason that it has now been translated into English. Those who know of it have already pointed to the benefit other scholars might derive from it for their studies of the British administrative and economic system in the eighteenth century,[6] and this English version, which includes minor updating and correction in the light of the debate since its first publication, should afford it the far wider dissemination that it deserves.

5 J. Macdonald, *Feeding Nelson's Navy* (London, Chatham Press, 2004), *The British Navy's Victualling Board, 1793–1815: Management competence and incompetence* (Woodbridge, Boydell, 2010); R. Knight and M. Wilcox, *Sustaining the Fleet, 1793–1815: War, the British Navy and the contractor state* (Woodbridge, Boydell, 2010); R. Morriss, 'Colonisation, conquest and the supply of food and transport: the reorganisation of logistics management, 1780–1795', *War in History*, 13 (2007), 310–24; J. Davey, 'Within hostile shores. Victualling the Royal Navy in European waters during the French Revolutionary and Napoleonic Wars', *International Journal of Maritime History*, 23 (2009), 241–60.

6 E.g. D. Baugh, *American Historical Review*, 117 (2012), 604.

Acknowledgements

I should like to take this opportunity to express my debt to all those who helped and guided me in my research. Above all, I wish to thank Professor Jean Meyer to whom I owe my interest in maritime history. Since 1982, he has ceaselessly, untiringly and attentively trained me in my research. I would like to offer him here my very respectful and affectionate gratitude.

I would also like to thank Professors Marine Acerra, Paul Butel, Pierre Chaunu, Philippe Haudrère, Philippe Masson, Etienne Taillemitte, Patrick Villiers, Michel Vergé-Franceschi and André Zysberg for their encouragement, warmth and kindness throughout my research.

I should also like to mention the great benefit I obtained from the advice and opinions of Professor Daniel A. Baugh of Cornell University, a specialist in British naval administration, and from Dr N.A.M. Rodger of Oxford University. I thank them both for their countless meetings with me and for our regular correspondence. I hope that the contents of this book will adequately express the gratitude I feel towards them.

This book was first published by Honoré Champion in Paris in 1999 as *Marine, économie et société. Un exemple d'interaction: l'avitaillement de la Royal Navy durant la guerre de sept ans*. In producing this English edition my thanks are due to the translating skills of the late Dr Anita Higgie of the Institut Catholique de Paris, and to Dr Michael Duffy of the University of Exeter for his assistance in completing the translation and in helping to update and make minor changes from the original text.

I am very grateful for the support of the Centre d'Etudes de la Mer of the Institut Catholique de Paris (CETMER), the Laboratoire d'Histoire et d'Archéologie Maritime (Université Paris IV-Sorbonne/Musée de la Marine), with special thanks to its Director, Professor Olivier Chaline, the Direction de la Mémoire du Patrimoine and the Archives de Ministère français de la Defence, especially M. Laurent Veyssiere. Thanks are also due for the support and encouragement of Peter Sowden of Boydell & Brewer.

Above all, my thanks go to the former President of Paris IV-Sorbonne, Professor Jean-Pierre Poussou. His numerous comments and suggestions, his faithful support and good-natured availability permitted me to successfully complete this book.

Introduction

> In spite of all its other weaknesses, European administrative organisation was one of Western Europe's greatest strengths during its conquering phase. This organisation allowed Europe to be one step ahead of the other continents. The ensuing result – the divide between Europeans and 'Others', was due to a large extent to the triumph of State and administrative organisation, in itself a product of the Industrial Revolution. Not to be forgotten, however, was also the role played by liberty …
>
> Jean Meyer, *Le poids de l'Etat* (Paris, PUF, 1983), 289

Professor Jean Meyer's opinion (above) can be aptly applied to British naval history in the modern period. It was thanks to the Royal Navy's ability to create a structured administration and thus stimulate 'cisatlantic' commercial capitalism (to use D.A. Farnie's[1] expression) which was, at the time, in a period of full development, that Great Britain was enabled to progressively dominate the high seas.

The author's study of the struggle during the seventeenth and eighteenth centuries to control the Caribbean and Atlantic coasts of Central and South America,[2] a veritable yardstick by which to measure the level of organisation of the various European navies, has already shed light on the fact that in order to succeed in a far-distant theatre of operations, it was indispensable that the means available should match the strategic ideas. Basically, there were two reasons as to why, towards the end of the Seven Years War, Britain was finally able to paralyse Spain's and France's rich trade with the Caribbean and seize their major defensive strongholds. First, Britain succeeded in complementing the idea of forward staging posts with that of an advanced source of supply by leaning more and more heavily on her North American colonies, and second, she endowed her Navy with a remarkably efficient administrative system. This twofold revolution also provided a solution to the problem of sickness, a remedy for which was indispensable for the successful outcome of long-distance operations.

1 D.A. Farnie, 'The commercial empire of the Atlantic 1607–1783', *Economic History Review*, second series, vol. XV, no. 2 (1962), 205–18.

2 Ch. Buchet, *La lutte pour l'Espace Caraïbe et la Façade Atlantique de l'Amérique Centrale et du Sud, 1672–1763* (Paris, Librairie de l'Inde Editeur, 1991), 2 vols, 1,297 pp.

The sheer importance of this last factor would be incomprehensible without bearing in mind the problems that resulted from the extreme length of voyages to the West Indies. Since the ships sailed in fleets, their speed was limited by that of the slowest ship, and the two to two-and-a-half months that it took to cross the ocean had a significant effect on the health of the crew. Death and sickness rates within squadrons arriving in the Caribbean ranged from 0.43 per cent to 17.2 per cent for mortality and from 5.39 per cent to over 32 per cent for morbidity. There were also indirect effects resulting from the length of these geo-strategic expeditions, coupled with the extremely insalubrious living conditions on board, which lowered the crews' resistance to the endemic illnesses specific to the Caribbean, in particular, the sadly notorious yellow fever.

The principal medical reason for such high rates of death and sickness was scurvy (vitamin C deficiency), which became especially problematic in voyages lasting over seventy-five days. This extremely debilitating affliction, which weakened their immune system, made sailors particularly vulnerable to 'putrid fevers' (better known today as typhoid fever) and to the 'malignant pestilential fever' (typhus). Also to blame were the poorly salted foodstuffs, particularly badly preserved during the first years of the War of the Spanish Succession; the putrid water; the dampness that permeated the vessels; the oxidisation of copper; the unsanitary conditions; and the hordes of rats that scurried amongst the tattered possessions of the sailors. All of these factors explained the outbreaks of typhus and typhoid fever on board vessels during the seventeenth and eighteenth centuries. Not to be forgotten in this dismal list of living conditions was the extreme overcrowding on board the ships. This was due to the fact that most of the expeditions carried troops on board, either in preparation for new offensives, or simply to reinforce garrisons already present in the West Indies. This overcrowding simply facilitated the spread of contagious diseases.

Statistical studies of the crew muster lists of the period preserved in the British Public Record Office and the French Archives Nationales have revealed that, generally speaking, British squadrons had much higher sickness and mortality rates in their ocean crossings than French expeditions.[3] By impeding the offensive power of the British squadrons, these deficiencies contributed to the protection of French and Spanish possessions in the Caribbean.[4]

The men consequently arrived at their destination in an extremely debilitated state and were thus much less resistant to the endemic diseases which

3 Buchet, *La lutte*, vol. II, 695–727.

4 Buchet, *La lutte*, vol. I.

threatened them from innumerable sources. Added to this was the fact that the main strategic sites were on the coast and surrounded by swamps, a terrain which was particularly hostile to the 'healthy well-being of the men'. Last but not least, most of the military operations were carried out during the rainy season. In fact, for both budgetary reasons and in order to avoid storms in the Bay of Biscay, the squadrons were usually only ready to depart by the beginning of the spring. The duration of the crossing scarcely left enough time for the expeditions to carry out their assignments before the wet season began in July, and the most terrible sickness disasters would increase with the duration of the expeditions. It was not unusual to find death rates soaring beyond the 50 per cent mark. In fact, the disease-related mortality rate rose to 61 per cent in the 1695 expedition against St Domingo led by Admiral Wilmot.[5]

Confronted, therefore, by the sheer scale of the phenomenon, and in an effort to contain the factors leading to the epidemics, various Admiralties rapidly undertook the following measures: first, systematically introducing a stop-over period at intermediate staging-posts so that the squadrons could recuperate both before and after operations, and second, carrying out operations as quickly as possible. Admiral Vernon estimated that the duration of an operation should not exceed six weeks.

However, for a variety of reasons, for a long time these two remedies proved to be irreconcilable, thus explaining the long chain of sickness disasters that continued to strike the Royal Navy. In fact, stopping-over at assembly and recovery bases in the Caribbean area delayed the start of operations, which then often destroyed the surprise factor that was so crucial to their being carried out quickly and successfully. For obvious reasons, the presence of a squadron in the region was not easy to hide, and the stop-over period often allowed sufficient time for the enemy to become aware of the intended target.

Furthermore, there was another paradox. In the course of time, both the West Indian islands and the Spanish-held continental mainland underwent a large increase of population. In order for the British to gain a foothold and maintain a presence, it was consequently necessary to increase troop numbers on the expeditions. It took longer and longer to raise and equip all these men, which again nullified the positive advantages of the surprise factor. Last but not least, it must not be forgotten that the ensuing delays resulted, more often than not, in the expeditions failing to reach the theatre of operations before the wet season, once again resulting in an increase in mortality rates.

In conclusion, the increase in the number of men, which was deemed necessary in order to bring operations to a speedy success, resulted in a signifi-

5 A statistic derived from the following sources: *Calendar of State Papers Colonial, America and West Indies*, vol. 14 (*1693–1696*), nos. 1983, 1946, 2026 and 2028.

cant decrease in the health of the troops, which in turn resulted in the fact that the British simply could not mount their expeditions in the time-frame available to them. Furthermore, the increase in the number of land forces on these expeditions led to autonomous commanders being appointed for the two separate arms, Army and Navy, and differences of opinion between the two proved to be a further impediment to acting speedily and thus successfully.

By moving to a more globally geo-strategic framework, and, as a corollary to this, providing its Navy with a better-structured organisation, Britain began progressively to break out from these constraints, and so succeeded in devising a system of intercontinental management for this type of expedition. Both in terms of recruitment and in terms of logistics, the British began to make more and more use of North America, as well as of Britain itself. This development began in 1740, when over 3,600 men were raised in the northern colonies to augment the troops under Vice-Admiral Vernon and General Wentwoth. The Duke of Newcastle personally requested that these men be treated with the utmost benevolence, so that in the future they would be inclined to participate in other such expeditions, and perhaps even to settle the lands that the British were hoping to conquer. Likewise, many of the food provisions and other essential commodities came from the thirteen colonies. Furthermore, Vice-Admiral Vernon did not hesitate to exempt the crews of all ships coming from these colonies from impressment, in order to encourage the men to participate.

During the Seven Years War, North America found that thanks to its proximity to the West Indies and to its now permanent relations with the staging-posts in the Caribbean, it had become a veritable advanced resource base, indispensable to the success of this type of geo-strategic expedition. Growing contributions from the northern colonies both in men and in food provisions would finally alleviate health problems, which had formerly crippled the expeditions. In fact, the 14,000 soldiers brought from North America in the Seven Years War[6] were a lot less exhausted by their journey than those coming from Europe. Furthermore, the dispatch of men and supplies from two distinct geographical points (North America and Europe) significantly reduced the quantity that needed to be assembled and deployed from any one point and thus had a direct effect in hastening departures. This organisation

6 Half of this contingent consisted of regular troops, the other half consisted of militia and of '*provincials*' (colonial volunteers who had engaged themselves for a particular task). PRO, CO 5/61, f.530. Recapitulation of troops from North America sent to the West Indies. T. Mante, *The History of the War in North America and the Islands of the West Indies* (London, W. Strahan, 1772); D. Syrett, 'American provincials and the Havana campaign of 1762', *New York History*, 49 (Oct. 1968), 375–90.

thus allowed the expeditions commanded by Moore and Hopson in 1759 and Rodney and Monckton in 1761 to carry out their assignments before the start of the wet season.

The local provision of stores for these expeditions, from the Northern American colonies, also had a beneficial effect on the health of the men. The geographical proximity of the North American continent, together with a shorter, more regular service by convoys (according to the Admiralty these convoys faced less danger, which encouraged the transporting merchant ships to sail these routes, as opposed to the reluctance manifested by merchant ships coming from Europe), allowed the crews to consume much fresher produce. However, it has to be remembered that it was only due to the demographic and economic development of the North American colonies that manpower, provisions and naval equipment were able to pass into the Caribbean area in the first place. Study of the logistics of food produce has shown that it would have been very unlikely that, on her own, Britain would have been in a position to supply the constantly growing forces that she was sending to this theatre of operations.[7]

Another positive offshoot of this coherent geo-strategic development is that the Admiralty began to run the Navy along highly structured lines, which proved to be the second factor leading to British success by the end of the Seven Years War. First of all, the command structure was modified. The freedom of operation once experienced by expedition commanders was now limited, to the Admiralty's advantage. Until 1759, commanders even had the freedom to choose their own objectives; however, Moore, Rodney and Pocock, in their respective campaigns of 1759, 1761 and 1762, found themselves issued with highly detailed operational plans which they had to follow closely. Furthermore, the functions and responsibilities relevant to the Navy and the Army respectively were also clearly drawn up. The treatises on amphibious warfare written by Thomas Molyneux and by John MacIntyre are witness to such developments.[8] In this way, central organisation was replaced by the idea of integrated command and at the same time was able to accommodate the parallel hierarchy of the two forces. Another direct and fortuitous consequence was that differences of opinion, which in the past had often been frequent amongst the main leaders of an expedition, now became the exception.

There was another consideration which should be taken into account when explaining the victories finally won by Britain.[9] Apart from an increase

7 Buchet, *La lutte*, vol. II, 905–24.

8 Thomas Crisp Molyneux, *Conjunct Expeditions: or expeditions that have been carried out jointly by the army and navy ...* (London, 1759); John MacIntyre, *A Military Treatise on the discipline of the Marine Forces* (London, 1763).

9 Buchet, *La lutte*, vol.. II, 590–691.

in firepower of the ships taking part in this type of expedition, there were major advances (far superior to what was happening in France) in the standardisation of naval construction, as can be seen by studying the lists of ships involved. All of these factors had a positive effect. In this way, by the start of the so-called 'War of Jenkins Ear' in 1739, a certain homogeneity had been reached regarding the vessels which participated in a particular expedition, and this greatly facilitated logistics in terms of naval stores and ship repair in the West Indies. Also not to be forgotten were the climatic conditions. They had always been particularly difficult in the Caribbean area, but this 'new' generation of vessels was far better equipped to cope with the weather.

Progress and innovation had taken place in the technology of hull-structure. In fact, a study of length and width ratios is very revealing: the average ratio for the 115 ships sent to the West Indies during the 1739–48 war and during the Seven Years War was 3.5682; whereas for the 92 vessels listed as being in service during the earlier War of the Grand Alliance (War of the League of Augsburg) and the War of the Spanish Succession, the average ratio was 3.7789. This evolution reflects technological choice: British vessels, unlike their French counterparts, had acquired bluffer, less streamlined hulls which, although less speedy, were much more robust. Consequently, during the war of 1739–48 and during the Seven Years War, British ships were far better adapted to the long periods of service that they had to undergo in the West Indies in order to maintain a large, permanent naval presence to gain and keep control of the area. Since in this region of the globe it was highly advisable to protect hulls from prolific sea worms, the number of 'doubled' ships (with hulls sheathed with an extra layer of planking) within the British fleet sent to the Caribbean increased – even before the Lords of the Admiralty made this a mandatory feature on 10 September 1761. As a consequence, there was now a range of appropriately equipped ships that could rapidly be deployed to the West Indies, chosen especially to improve the fleet serving in this theatre.

Another example of the extraordinary degree of organisation which helped the British Navy to surmount previous obstacles was the development of signals. These assisted the maintenance of a cohesive convoy formation amongst the ships of the expedition, and meant that fleets could be sailed across dangerous zones and thus gain valuable time. For example, Pocock's powerful expedition in 1762 now had the means to sail down the 'old Bahama Channel' *en route* to Havana, thus granting the expedition the advantage of surprise. Since at that time the route was considered to be impassable, Spanish commanders only anticipated an expedition – even a small one – reaching Havana via Cape Florida or Cape San Antonio. Thus the British fleet was able to reach the town without the alarm being raised by Spanish ships.

Disembarkation procedures also benefited directly from the development of signals: the troops were able to land in perfect order, in the positions they normally held in combat formation, thanks to the mooring arrangements of the landing craft. The Navy was thus enabled to increase troop numbers in the first assault without putting their military value at risk. The use of flat-bottomed boats, developed for this type of operation early in the Seven Years War, also allowed for far more rapid disembarkation, which in turn meant that muskets were spared prolonged contact with sea water and their gunpowder and flints were less likely to get wet.[10]

The relative ease with which victualling was carried out for the squadrons in the West Indies[11] is possibly one of the best yardsticks by which to measure the advances made by the Royal Navy in this period. In other words, the development of the Victualling Board's outposts in the West Indies, the improved dialogue between the Commissioners of the Board and the Lords of the Admiralty during the Seven Years War, the huge increase in suppliers with whom the Board did business and their increased specialisation in terms of product and/or geographical area, and the constant pressure applied by the Commissioners on their suppliers by their encouragement of competition, along with the growing importance of the supply role played by the North American colonies, were all factors which help to explain how Britain succeeded in satisfying the increasing needs of her constantly growing naval forces which were being deployed to the West Indies.

Inasmuch as health and logistics proved to have such a direct influence on the success or failure of operations in the period under consideration, this book will focus especially on these issues. Having been struck by the markedly higher mortality rates suffered by British fleets deployed to the West Indies, compared to a much lower rate for French fleets, and having observed that this difference was marked by factors occurring during the transatlantic crossings, the author explored these factors, since they were so powerful in seriously hampering the British Navy's offensive potential for the better part of a century. A thorough examination was made of the Europe–West Indies routes.[12] Lengthy research based on British and French log books threw up some surprising and interesting discoveries, many of which have proved to run counter to some commonly held theories, for example on the question of sea routes and their duration, which appear, in fact, to have been significantly different for British and French fleets. It transpires that while all the French

10 See D. Syrett, 'The British landing at Havana: an example of an eighteenth century combined operation', *Mariner's Mirror*, 55 (August 1969), 325–31; Syrett, 'The methodology of British amphibious operations during the Seven Years and American Wars', *Mariner's Mirror*, 58 (August 1972), 269–80.

11 Buchet, *La lutte*, vol. II, 820–1098.

12 Ch. Buchet, 'Des routes maritimes Europe-Antilles et de leurs incidences sur la rivalité franco-britannique', *Histoire, Economie et Société*, vol. 13, no. 4 (1994), 563–82.

squadrons passed exactly half-way between the Azores and Madeira in order to reach the West Indies, the British expeditions all took a slightly more southerly route in order to put in at Madeira. The French Navy included wine in its rations, the Royal Navy carried beer. However, since beer did not keep beyond three weeks at sea and deteriorated quickly in warmer climates, it was necessary for the British to stop *en route* to procure reserves of wine. It has been possible to recover most of the delivery slips attesting to wine delivery in the ADM 110 series of the Public Record Office.[13] When these detours are added to time spent anchored at Madeira and getting ready to set sail again, and, furthermore, when it is taken into account that setting sail from Britain in the first place took much longer for the British squadrons than it did for their French counterparts (due to the far less favourable position of the English Channel ports when compared to those of the French Atlantic coast, as far as the prevailing winds were concerned), then it is not surprising that the British voyages lasted substantially longer than the French ones.

It has also been of interest to investigate the utilisation of press-gangs for the rounding up of men when the Royal Navy needed to increase its numbers. This practice led to the early appearance, again in comparison to the French Navy, of scurvy within British ranks due to the fact that the men were crammed into receiving hulks for at least two months before being assigned to a ship. The press-gang system when applied in the West Indies also caused other problems: it weakened the colonies' defensive system, and was at the heart of serious tensions between the colonists and the Navy.[14]

The administrative, economic and logistical aspects studied in the author's doctoral thesis showed that the victualling of an increasingly larger fleet sent by Britain to the Caribbean came to be carried out more and more efficiently and cost-effectively. This was followed up by an examination of global logistical issues to better understand this development and the degree of efficiency of the naval administration. Scrutiny of the Navy Board's registers (ADM 106 series) recording naval equipment confirmed the extraordinary improvement in organisational development by the Royal Navy.[15] This development was in stark contrast to that of the French fleets, also analysed.[16] Another comparative study, on the supply management of ordnance equipment

13 The names of the various suppliers as well as their prices are listed in the above article.

14 Ch. Buchet, 'Les levées d'hommes opérées par la Royal Navy aux Antilles entre 1689 et 1730', *Histoire, Economie et Société*, vol. 10, no. 4 (1990), 521–43.

15 Ch. Buchet, 'Les modalités évolutives de la logistique anglaise en matériel naval dans l'espace caraïbe (1689–1783)', *Histoire, Economie et Société*, vol. 11, no. 4 (1992), 571–96.

16 Ch. Buchet, 'La logistique française en matériel naval dans l'espace caraïbe (1672–1763)', Comité de Documentation Historique de la Marine, Communication de 14 Janvier 1992 (Vincennes, Service Historique de la Marine, 1993), 43–74.

(cannon, cannon-balls, gunpowder), which depended on a separate specialist organisation – the Ordnance Board – completed this earlier survey of the logistical systems implemented overseas.[17] Since the Commissioners of this Board were not under the direct authority of the Lords of the Admiralty, they did not hesitate to be totally direct and sometimes even critical of the Admiralty in their correspondence. Analysis of these documents has enabled re-evaluation of the means by which the Lords of the Admiralty carried out their tasks of stimulating and co-ordinating so as to better judge the results obtained by the Victualling Board and the Navy Board.

The parallel development of naval bases also had to be considered, since the astonishing development of their arsenals in the West Indies was an excellent indicator of Britain's expanding colonial determination.[18] It has been possible to monitor this development by examining accounts in documents concerning personnel located in the ADM 42 series.

With this and improvements in the logistical system for naval equipment, the British squadrons became more and more effective. The immobilisation of half a squadron's ships while repairs or careening were carried out was no longer necessary during the Seven Years War. The needs of the squadrons in terms of, for example, food provisions and naval equipment, were undertaken more efficiently and at less cost. Without a doubt, the determining factor for British success during the Seven Years War was the development of naval administration and naval infrastructure which contributed to the decrease in cost per capita of the forces sent out to this area.

Such research into these distant expeditions has been highly revealing of the degree of organisation of the different European navies and has led us to the modes of central operation, key factors in the success or failure of these long-distance operations. Basically, we have shown that the administrative development of the Royal Navy was the principal element which allowed Britain to gain dominance of the Caribbean, and having studied the logistical systems between the British and French homelands and their overseas territories, it seemed that the next logical and necessary step was to extend the field of investigation and analyse, further back in the chain, the victualling of the Royal Navy – how it functioned, and its economic effects.

Looking beyond the maritime context, it seemed that this field of study could also, by a systematic investigation which reconstituted the body of

17 Ch. Buchet, 'L'Ordnance Board et la logistique en armes pour les flottes anglaises aux Antilles (1689–1763)', *Mélanges offerts à Monsieur le Professeur Jean Meyer: Etat, Marine et Société*, ed. M. Acerra, J.-P. Poussou, M. Vergé-Franceschi and A. Zysberg (Paris, Presses de l'Université de Paris-Sorbonne, 1995), 73–87.

18 Ch. Buchet, 'Révélateurs d'une détermination coloniale: les arsenaux anglais de l'espace caraïbe', *Histoire, Economie et Société*, vol. 16, no. 1 (1997), 123–49.

merchants and wholesalers engaged in contracting with the Victualling Board, contribute to the work already begun by Professors F.J. Fisher and F. Crouzet, to reveal the degree of overlap with the agricultural and commercial parts of regional economies, and thus evaluate the interactions that existed between navies and economies. That this study can be carried out is thanks to the fact that sources found at the Public Record Office of the British National Archives and at the Corporation of London Record Office are more complete and less heterogeneous than their counterparts at the French Archives Nationales. Did the development of the Royal Navy, with its ensuing effects on commerce and thus on finance, provide a link between the agricultural revolution and the industrial revolution, consequently contributing to the economic primacy of Great Britain up to the end of the modern epoch?

It would then be interesting to compare the British system of victualling under the aegis of the Victualling Board with the French system established by Colbert in 1669 – the office of Victualler-General of the Navy (*Munitionnaire Général des Armées Navales*). The French study would lead to a series of regional surveys, which would be indispensable in evaluating the manipulation of markets as an instrument of force wielded by the state. It would then also be possible to evaluate the real and/or virtual effects that the Navy had on the development of the French economy.

The Seven Years War was the obvious choice for this study: first, because it was during this period of hostilities that the British Navy reached its peak in terms of logistical advancement before the independence of the thirteen North American colonies reduced for a time the world's largest free market zone; second, because of the opportunity to situate this survey chronologically with the remarkable work carried out by Professor Daniel A. Baugh on the administration of the British Navy;[19] and finally, the Seven Years War is short of specialised maritime in-depth studies, with the notable exceptions of N.A.M. Rodger's work on the war's social aspects, that of S.F. Gradish on recruitment, and that of D. Syrett on the provision of shipping to carry troops and supplies.[20] This study will present the organisational aspects of supply that were gradually put into place, the bases, and finally the principal commodity markets.

19 D.A. Baugh, *British Naval Administration in the Age of Walpole* (Princeton, Princeton University Press, 1965), *Naval Administration 1715–1750* (London, Navy Records Society, 1977), vol. 120.

20 N.A.M. Rodger, *The Wooden World* (London, Collins, 1986); S.F. Gradish, *The Manning of the British Navy during the Seven Years War* (London, RHS, 1980), though errors in this work were uncorrected owing to the premature death of its author; D. Syrett, *Shipping and Military Power in the Seven Years War* (Exeter, Exeter University Press, 2008).

This study begins by seeking to elucidate how, between 1546 and 1756, the Admiralty succeeded in reconciling the advantages of a private supply system with the imperatives of a state-run task and created an organisation that successfully blended flexibility with steadiness of purpose. This was worked out in a controversy concerning the victualling of the Royal Navy which, from the seventeenth century, opposed the supporters of state intervention, who highlighted the contradiction inherent in a system which claimed both financial profit and the common good as its aims, and the partisans of a private system who preferred to put forward the financial economy of a competitive contractual system based on an invitation to tender bids. These issues form the basis for Chapter 1 of this book.

The next step, in Chapter 2, reconstructs how this organisation operated under the aegis of the Victualling Board during the period of the Seven Years War. This will involve a detailed analysis, including a financial one based on the ledgers in the ADM 20 series in the Public Records Office. These documents allow us to evaluate what effect late payment had on rates of discount, and also to evaluate the relative difficulty the Crown had in financing this expenditure. This in turn will throw light on the financial capacity of the country to continue, if not increase, the war effort.

In Chapter 3, the final organisational aspects will be considered. The chapter will focus on the food rations and their evolution, and on the technical aspects of food preservation. The benefits gained from scurvy prevention will be evaluated, as will the medical knowledge that was available at the time on vitamin-deficiency-related sickness. The direct or indirect effects that the Seven Years War had on an understanding of this malady will be quantified.

In the second part of the book, Chapter 4 will deal with naval bases, and in particular with the management of victualling warships at the six principal yards which fell under the direct jurisdiction of the Victualling Board: in order of size, London, Portsmouth, Plymouth, Chatham, Dover and Gibraltar. It will examine the evolution of their facilities from a geo-strategic point of view, and their respective importance in terms of numbers employed. This data is found in the ADM 113 series which holds documents relating to labour.

Chapter 5 offers a micro-economic examination of the day-to-day running of Plymouth, which was selected because, apart from London, it was the most active port in terms of victualling in the latter years of the Seven Years War. Furthermore, this base was particularly worthy of study in that it succeeded in victualling an ever-increasing number of squadrons despite its incomplete facilities.

Chapters 6 and 7 will then examine the victualling at ten secondary home ports and eighteen outlying stations by merchants under contract to the Victualling Board. A geographical breakdown of these suppliers offers an insight into the networks operated by the major merchant houses who dominated transatlantic commerce.

The last section sets out to identify and analyse the merchant body engaged in the process of victualling the Navy, according to market sectors. This data was compiled from a thorough scrutiny of the ADM 111 series which lists the names of all the Royal Navy's suppliers and also their unsuccessful competitors. By grouping and re-grouping these lists according to criteria such as names of suppliers according to season, or merchants' associations, it is possible to establish how far the contracts which were offered following an invitation to tender bids by the Victualling Board were or were not influenced by hidden agendas and understandings. Did the sheer quantities of provisions to be supplied not limit the conditions of access to these markets, thus threatening an indispensable diversity of supply? How could the administrators of the Victualling Board based in London conduct almost every purchase?

Finally, Chapters 8, 9 and 10 focus on commodity market studies, which lead to a greater understanding of the fluctuations in price relative to quantity, and consequently clarify developments in the cost per man of victualling.

PART ONE

THE GENERAL ORGANISATION OF VICTUALLING THE BRITISH NAVY

1

From an Empirically-Based Organisation to a Centrally Planned System: The Strengths and Weaknesses of the Victualling Board

In 1659 John Hollond wrote, 'It hath been an old and great dispute which is the best way of victualling, whether by particular men as contractors at a certain rate, or by the State.'[1]

Throughout the seventeenth century a fierce debate raged concerning the strategy of victualling the Royal Navy, and those in favour of a competitive private system were already clashing with supporters of state intervention. The former group argued in favour of a system that offered a certain degree of flexibility, a dynamic approach and lower costs due to a contractual system based on market demand; whereas the latter group, which included people such as John Hollond, underlined the contradiction inherent in the desire of the former to make a profit on a mission of public service. Hollond claimed that, in the same way that the presence of a worm can rot a piece of fruit, corruption would be the inevitable consequence of a victualling system awarded to merchants and wholesalers. He concluded by stating that the quest for profit making was apparent at every level and incited officers and suppliers to reach 'agreements', to the detriment of the public interest.[2]

It may be argued that the Royal Navy's excellent logistical results, starting with the war of 1739–48 and culminating in the Seven Years War, were due to the success the Admiralty achieved, after a certain experimental period, in reconciling the flexibility and rigidity of the two systems outlined above into a single, original, organisation.

The Experimental Period (1546–1715)

It was under the reign of Henry VII, the first Tudor, that England, having just emerged from a long civil war, first turned towards modernity. Henry,

1 J.R. Tanner, ed., *Two Discourses of the Navy 1638 and 1659 by J. Hollond* (London, Navy Records Society, 1896), vol. 7, 154.

2 Tanner, *Two Discourses … by J. Hollond*, 152–67.

who had conquered Richard III and was a contemporary of the discovery of America, realised that this geographical revolution was to have far-reaching consequences for England. In fact, the ports of call on the return sea-routes from America conspired to throw England into the heart of the New World. The expansion towards the east that had led to the Hundred Years War was now replaced by expansion towards the west, and by 1497, John Cabot, sailing under the king's protection, reached the North American coast.[3] Raleigh, Drake, Frobisher and Cavendish, thanks to their successes, highlighted England's maritime calling, while Sir John Hawkins at the Navy Board began to draw up the administrative outlines of the Royal Navy.[4]

As far as victualling was concerned, the Admiralty placed itself in the hands of one or two important merchants, better known as contractors. This system was maintained for almost two centuries. Having decided on a price in advance, the contractor had to deliver a set number of rations – decided upon by the state and which the king's council proclaimed each year at the beginning of autumn – to the country's major ports. The contractor also had to be prepared to satisfy any extra orders issued by any of the Navy's authorities.

This system was in fact remarkably similar to the one which had been adopted by France in October 1669 on the orders of Colbert, and which called for the setting up of a 'Victualler General of the Navy' – a private enterprise which was responsible for the Navy's victualling needs, regardless of the geographical location of the port.[5]

In fact, this system did not prove any more efficient in France than it did in England, and poor victualling severely compromised the offensive potential of the Navy. Numerous examples of this abound throughout the sixteenth and seventeenth centuries,[6] beginning with the English fleet that confronted the Spanish Armada in 1588 and was the victim of abnormally high mortality rates due simply to defective victualling. The naval authorities in the Tudor and Stuart periods advised that the management of victuals was 'more trouble

3 John Cabot named this land *Terra Prima Vista*. But was it, in fact, referring to Newfoundland, Labrador, Nova Scotia or the Maine coast? See M. Duchein, 'Canada: deux découvreurs pour un pays', *L'Histoire*, 146 (July–August 1991).

4 Created in 1546 under Henry VIII, the Navy Board was originally responsible for all administrative concerns within the Navy. However, in light of the scale of the task, its original role was progressively restricted, and the Victualling Board was inaugurated in 1683. In the eighteenth century, the thirteen Commissioners who were in charge of the Navy Board managed, in particular, all naval construction, the dockyards and logistics of naval material. N.A.M. Rodger, *The Admiralty* (Lavenham, Terence Dalton, 1979); Buchet, 'Les modalités évolutives de la logistique anglaise en matériel naval dans l'espace caraïbe'; B. Pool, *Navy Board Contracts 1660–1832* (London, Longman, 1966).

5 Buchet, *La lutte*, vol. II, 925–61.

6 Cf. the references given by Baugh in *British Naval Administration*, 373, n.2.

than any of the other details of administration'.[7] There were any number of factors which, when combined, explained why the system could not function and highlighted the difficulties involved in rectifying it. These factors included the sheer scale of the task facing the contractor; the Admiralty's inability to annul the contractor's contract should a legitimate reason arise, without detriment to the victualling system; and a precarious budget, which had to be managed by successive governments.

A first opening was nonetheless created in 1654, during the religious and political revolution which was agitating England at the dawn of Cromwell's 'Western Design' and which allowed England to take Jamaica. What happened is that for the first time the state took direct control of the management of victuals for the Navy. Crudely carried out, the operation was nothing other than an improvisation without resources and did not result in any of the positive effects that had been counted on.[8] However, the onset of the Restoration ended the experiment, and in 1660 the old system was once again introduced. However, this brief interlude must not be dismissed out of hand, for it signalled a change in attitudes that was beginning to be heard.

In 1679, having just changed suppliers, the Lords of the Admiralty created a new position, that of Surveyor of Victuals, whose role was to ensure that the contractor did not cut corners.[9] It was along these lines that the Victualling Board would be created. The victualling of the Navy was therefore still within the domain of private enterprise, but under the tight control of the state. Within four years, in order to render the post even more efficient, the office of Surveyor of Victuals was abolished, and in its place was created a Board dependent on the Navy Board – the Victualling Board. It was composed of seven salaried Commissioners who were under the direct control of the Admiralty. As of 1 January 1684, the Commissioners took direct control of the Navy's victualling needs, and tradesmen and merchants were now only called upon for specific, highly selective needs as and when the occasion arose. However, due to a long list of negative factors such as lack of experience; the need to rethink and rework a whole new system and the deplorable state of the economy following the Revolution in 1688, it is hardly surprising that the early Commissioners failed in their task.

When the War of the Grand Alliance (1689–97) – also known as the War of the League of Augsburg – broke out, the Commissioners found themselves unable to victual a Navy that was actively at war. The House of

7 M. Oppenheim, *A History of the Administration of the Royal Navy … from MDIX to MDCLX* (London, J. Lane, 1896), 81.

8 Oppenheim, *A History of the Administration of the Royal Navy*, 325–7.

9 J.R. Tanner, *A Descriptive Catalogue of the Naval Manuscripts in the Pepysian Library at Magdalene College* (London, Navy Records Society, 1903), vol. 26, I, 153–7.

Commons heard of the problem and consequently, in November 1689, asked the Commissioners to explain themselves. Parliament then terminated the contracts of a number of Commissioners and placed Thomas Papillon, a man of experience, at the head of the Victualling Board.

The new management of the Victualling Board was, however, equally unconvincing. Josiah Burchett, who was a Navy administrator during the reign of William III, wrote, 'The want of timely supplies [of victuals] ... at several times ... greatly obstructed those designs, which otherwise might have been carried out on greater advantages to the Publick.'[10] The Commissioners' goal was to acquire essential victuals, but at the same time they lacked the necessary funding. Rather than set up a rational and reliable logistical system based on clearly established procedures in terms of the purchase, storage and distribution of victuals, they were forced to operate speculatively, case by case, thus over the years destroying any semblance of organisation that the state had hoped to introduce. Far from being in a position in which they could place the merchants in direct competition with each other, the Commissioners were reduced to begging for credit from the very wholesalers that they were supposed to be controlling. Basically, this state organisation which was aimed at optimising the contractor system under the auspices of the Victualling Board, in the context of a mixed economy, required a financial capacity that was not available to seventeenth-century governments.

One disappointment after another soon dealt a serious blow to morale. As soon as the country found itself once again in a period of peace, criticism began to be directed against the naval administration, first by Parliament and then by the government. The administration was accused of bad management and even of corruption.[11] The Commissioners of the Victualling Board were dismissed, and a new team took over the Tower Hill site.[12] This marked the beginning of a period of intense development for the Victualling Board. At the instigation of the Lords of the Admiralty, the new Commissioners were faced with the task of setting up a sound and solid organisation that would no longer be at the mercy of the whims and fluctuations of the market. Past wrongdoings had to be righted, and this meant establishing regulations,

10 Burchett quoted by Baugh in *British Naval Administration*, 373.

11 R.D. Merriman, *Queen Anne's Navy* (London, Navy Records Society, 1961), vol. 103, 248.

12 Close to St Catherine's Dock, the Tower Hill buildings had been constructed for Navy Board and Victualling Board use on the site of the former St Mary Graces Abbey. Given over to purely Victualling Board use as of 1654, the complex consisted of administrative offices, warehouses, a cooperage, a bakery and a butcher. The Commissioners were also provided with lodgings in the complex. In 1701, the central corps was reconstructed.

constructing warehouses and creating the infrastructure that would bind all these elements together.

On 4 January 1701, the Admiralty presented the Commissioners with directives that had been revised, and now clarified a certain number of points.[13] The new team was horrified at the preceding administration's poor management. In 1703 the Commissioners published an audit in which they evoked the '*greatest confusion*' in relation to the accounts they had inherited. Arrears of £739,958/9s/11d had to be paid off without the slightest possibility of being able to verify the debts; credit slips which should have been signed by three Commissioners had only been signed by two; Victualling Board Agents in ports such as Chatham, Dover, Portsmouth, Plymouth and Harwich had rarely returned statements and accounts, and even when these did arrive they were not detailed, resulting in a total lack of verification and the possibility of fraudulent trafficking.[14]

The new Commissioners tried their utmost to remedy the situation. That same year, they ordered quarterly statements to be submitted by their Agents in which all operations could be clearly traced.[15] By 1706, all the Agents were well versed in procedures that ensured that all transactions were traceable and performed according to clearly defined regulations. By 1710 the Commissioners obtained from the Treasury the right to place a clerk of the cheque in each port, whose job was to check on the Agents and wholesalers.[16]

The administrative officers of the ensuing Victualling Board, who took over at the close of the Spanish War of Succession and in whose interests it would have been better to minimise the achievements of their predecessors in order to more greatly vaunt their own future accomplishments, were nonetheless forced to admit that a great deal of progress had been made by the retiring Board.[17] Furthermore, based on lessons which had been learned during the course of this war,[18] the Lords of the Admiralty issued two directives, in 1715[19] and in 1716,[20] which stipulated the precise tasks assigned to

13 National Maritime Museum, Sergison Collection, no. 85, 1–25. Merriman provides a large extract in *Queen Anne's Navy*, 254–8.

14 Commissioners of the Victualling Board to the Lord Treasurer (30 October 1703), in Merriman, *Queen Anne's Navy*, extract no. 9, 262–8.

15 Baugh, *British Naval Administration*, 449.

16 Victualling Board to the Admiralty, 22 February 1715, in Baugh, *British Naval Administration*, extract no. 1, 407–8.

17 Ibid.

18 Salted foods, in particular, had failed in the early years of the War of the Spanish Succession. Buchet, *La lutte*, vol. I.

19 PRO, ADM 7/648.

20 PRO, ADM 49/59. See also Appendix 6 below.

each and every member of the Victualling Board, and the procedures to be followed in relation to the carving up and salting of meat, food storage and its distribution. In addition to the instructions of 4 January 1701 already mentioned, these two texts now formed the core or corpus of the general rules which would guide the operation of the Victualling Board throughout the eighteenth century.

The Period of Development (1715–56)

Once the main rules had been laid down and the Commissioners had begun their struggle against fraudulent practices, the Victualling Board began a period of significant expansion. This growth occurred simultaneously in two directions: geographical expansion and structural development.

The victualling of Royal Navy ships in distant ports that did not possess a resident Victualling Board Agent left much to be desired. In these cases, the victualling task was left to the ship's captain, helped by the ship's purser, who would buy local produce which he paid for either in cash or (more usually) by issuing a bill of exchange drawn on the Victualling Board. This latter system was subject to much abuse and the captains and pursers often worked together to profit from it.[21] All that was required was to state that more articles had been bought than was the case, or to declare a higher price than that which had really been negotiated, in order to make a substantial profit.

In an attempt to remedy the situation, the Commissioners of the Victualling Board expanded their geographical area of intervention by creating new victualling yards under the management of their Agents, to whom all captains had to report in order to obtain additional victuals. This became the norm in Gibraltar, Port Mahon, Jamaica and Kinsale.[22] In 1715, the Commissioners decided that in areas where maritime traffic was not heavy enough to warrant the creation of a victualling yard, they would work with a private entrepreneur with whom prices had been agreed upon in advance, and who would be in a position to supply all the victuals that might be requested of him. Victualling contracts were drawn up in this way in Dublin,[23] then in Barbados, New England, New York, Virginia and Maryland.[24] A directive issued on 5 March 1716 by the Lords of the Admiralty stated that it was forbidden for captains

21 Baugh gives many examples of this in *British Naval Administration*, 391–4.

22 PRO, ADM 110/15 (28 January 1747).

23 Victualling Board to the Admiralty, 22 October 1716, in Baugh, *British Naval Administration*, extract no. 7, 415–16.

24 Victualling Board to the Admiralty, 12 November 1715 and 14 February 1718, in Baugh, *British Naval Administration*, extracts nos. 4 and 10, 412–13 and 418–19.

to re-stock their ships independently, and that all transactions had to pass through approved contractors.[25]

It took all the energy and determination of the Commissioners to ensure that these measures were carried out. The captains, deprived of what they had come, over time, to consider an essential supplement to their earnings, did their utmost to bypass the system. Many a contractor complained that the captains would buy from them only the more expensive products, and that they continued to buy other goods – those on which a profit could easily be made and hidden – elsewhere. Some captains did not even bother to go to a contractor's warehouse, claiming that the urgency of the situation required putting in at the closest port of call in order to stock up on provisions.[26] The Commissioners fought tooth and nail against this attitude, for contracts were at stake. Other similar attitudes engendered great 'discouragement to the service and very often a very heavy loss to the contractor, who is obliged to have a large stock on hand to answer any demand upon him'.[27] Slowly but surely, the decreasing number of complaints lodged by the contractors suggest that the severe attitude adopted by the naval authorities towards the recalcitrant captains eventually bore fruit.

This exacting system was pursued in 1738 when the Royal Council, at the request of the Lords of the Admiralty, decided that pursers should be appointed to all six rates of warship, and not, as was previously the case, only the first five; and that there should be a purser in every ship carrying a crew of sixty or more. This was in an attempt to limit the absence of control that had benefited the captains of these ships in victualling matters.[28]

Not only did the Victualling Board extend its geographical boundaries, it also extended its field of activities. At one end of the scale, the Commissioners decided that it was high time to centralise buying, and to this end they opted to make use of the London market as far as that was possible. The result was that they themselves established the contracts and became familiar with fluctuations in the markets.

At the other end of the scale, the Commissioners endeavoured to obtain the basic raw materials for foodstuffs which the Victualling Board itself would then process. In this way there was more efficient quality control of the food bought. The freshness and quality of peas, for example, could be more easily determined before cooking them. Similarly, quality control could

25 Victualling Board to the Admiralty, 14 February 1718, in Baugh, *British Naval Administration*, extract no. 10, 418–19.

26 Baugh, *British Naval Administration*, extract nos. 7, 10, 22, 31, 415–19, 429, 434–5.

27 Baugh, *British Naval Administration*, extract no. 22, 429.

28 Victualling Board to the Admiralty, 22 November 1728, in Baugh, *British Naval Administration*, extract no. 25, 430–2

be applied more easily to cereals than to flour. In this way, and over time, the Commissioners achieved the construction and development of essential infrastructures (this will be dealt with more specifically in a later chapter).[29] Breweries, bakeries, slaughter houses, salting centres, cooperages and warehouses in London, Portsmouth, Plymouth and to a lesser extent in Chatham and Dover began to grow into veritable victualling stations – Victualling Yards, a title imitating that of the Dockyards, the naval shipyards which were under the auspices of the Navy Board. A direct result of this evolution was the considerable increase in the number of permanent positions within the Victualling Board, as can be seen in Appendix 1.

The net result of these improvements was that victualling benefited from being far more efficient during the War of the Austrian Succession. For example, compared to the previous War of the Spanish Succession, there was less need for the implementation of *short allowance*, in which a sailor was authorised to receive only two-thirds of the prescribed daily rations, the remaining third being dispensed as cash. Also the fact that on many occasions the Commissioners were able to satisfy sudden and substantial demands equally attests to the Victualling Board's new-found efficiency.

The post-war development of these strategic foundations would result in dependable, highly effective victualling during the Seven Years War, despite the record number of men serving in the Royal Navy – 70,000 men on average, compared to approximately 40,000 men enlisted during the War of the Austrian Succession.[30]

29 Baugh, *British Naval Administration*, extract nos. 14, 18, 20, 27, 33, 39, 42, 422–3, 425–8, 435, 439–40, 442–4.

30 Baugh, *British Naval Administration*, 406; Merriman, *Queen Anne's Navy*, 251–2.

2

The Victualling Board and the Seven Years War

The management and efficient running of the Victualling Board was in the hands of seven central Commissioners who had been appointed by the Lords of the Admiralty, to whom they had to report two to three times a week.[1] The Commissioners usually met on Mondays, Wednesdays and Fridays.[2] Each Commissioner had clearly defined roles and responsibilities. Two Commissioners were responsible for finance and accounts, and the remaining five each had a particular field of responsibility across all geographical areas (see Appendix 2). They were aided in their tasks by a team of about forty assistants.[3] From the records available it has been possible to establish that, for instance, Mr Denham Skeet was responsible for making calculations concerning the suppliers' offers, and that a certain Mr Richard Henshaw was responsible for the victualling of geographically distant bases.

Victualling Methods

The total quantity of goods that the Commissioners needed to obtain to ensure the smooth running of victualling operations in any given year between 1 January and 31 December was estimated in the preceding autumn by the Lords of the Admiralty in a document entitled 'Declaration of victuals for the year x'. This document stated the needs for each major port so that the Commissioners could best distribute the victuals. Naturally, these figures were provisional and not 100 per cent accurate since they could only be based on the previous year's figures. In the event of war being declared, or the restoration of peace, a supplementary, modified inventory was issued.

In certain respects the document appears archaic, dating back to a time before the advent of the Victualling Board when victualling had been entrusted

1 Public Record Office (PRO) archives. ADM 110 and 111 series.

2 ADM 111 series.

3 This figure was deduced from the accounts records in ADM 20/225 (March 1757).

to a single contractor. In effect, it was a legal document, delegating all victualling responsibility to the contractor. The date on which it was issued was also of great importance, since the supplier was not obliged to commence his preparations until that date.[4]

In accordance with general instructions which required the Commissioners, if not to directly manage the whole operation, then at least to control it,[5] two different methods of victualling were simultaneously available to them.

The Victualling Board established a private supplier for each second-level home port and for each overseas port or station (see Table 2.1). These suppliers had to be ready to furnish all Royal Navy ships on request with the entire range of foodstuffs that made up a sailor's rations. Contracts issued in this category were irrevocable during the first year. They were valid until termination by either the supplier or the Victualling Board, with six months' notice having to be given.

Table 2.1 Ports and stations victualled by a private contractor

British secondary ports		
	Bristol Greenock Hull Leith Liverpool	Lynn Milford Newcastle Whitehaven Yarmouth
Outlying ports and stations		
Ireland	Belfast-Carrickfergus Cork Dublin	Londonderry Waterford
Intermediary bases	Oporto Lisbon	Madeira Senegal Goree
West Indies	Barbados – Leeward Islands Jamaica Havana	
North America	Louisburg Nova Scotia Quebec	New York South Carolina

4 Tanner, *A Descriptive Catalogue*, I, 171; Baugh, *British Naval Administration*, 386–9.

5 'The commissioner calling for branching out the whole business of their office, in such manner as that there might not be any part wherof but what should come under the particular inspection of some one Member of the Board' (General Instructions in PRO, ADM 7/648).

The question was repeatedly raised as to whether it would be in the best interest of these places for the Victualling Board to be directly responsible for their victualling, or whether the task should fall to a private contractor. It appears that the Commissioners themselves were of the opinion that it was more advantageous to make use of the latter when the number of ships being victualled was fairly small. This resulted in the Victualling Board no longer having to maintain warehouses, which would, in turn, have been placed under the supervision of a local Agent and his assistants, and their salaries would have been uneconomic if the number of rations they issued were low. Estimates showed that entrusting the victualling of the Royal Navy based at Jamaica to a private supplier would have resulted in savings of almost 14 per cent.[6]

On the other hand, victualling in the principal naval ports was too huge a task to be entrusted to a sole supplier. It seems that the Victualling Board's hands-on management did result in their domination of the market; better control of the quality of the foodstuffs; and a more efficient battle against fraudulent practices generated by the all-consuming desire for profit. In this way, the Victualling Board itself managed the victualling of the ports of London, Portsmouth, Plymouth, Chatham and Dover.

Rather than allow their Agents in these ports to draw up contracts with the suppliers of various products, the Victualling Board Commissioners had the authority to directly carry out this task themselves, which they usually tried to do.[7] This authority had been granted to them in 1704, following a survey carried out at Parliament's request which had brought to light certain instances of malpractice by some Agents. Consequently, almost all of the victualling was negotiated from London by the Commissioners.

This may seem surprising. When one considers the enormously varied range of foodstuffs the Navy required, not to mention the huge amounts of victuals required to feed 50,000 to 85,000 men,[8] it seems impossible that operations on such a scale could have been centralised at a single centre, albeit London. However, as the nation's capital, and by far its biggest port, manufacturing and financial centre, London had a population in 1750 of about 650,000 (11 per cent of the population of England and the biggest city in Europe).[9] As W. Maitland, F.J. Fisher, G.E. Fussell, R.B. Westerfield, C. Goodman

6 PRO, ADM.110/15, pp. 216–17. The Commissioners of the Victualling Board to the Lords of the Admiralty (8/01/1748).

7 PRO, ADM7/648, General Instructions.

8 Without taking into account the number of men that needed to be victualled in the colonies ('*Plantations*'). Cf. provisional figures in the ADM 111 series.

9 C. Chalkin, *The Rise of the English Town 1750–1850* (Cambridge, Cambridge University Press, 2001), 5–6.

and N. Gras have shown, London was the heart of the English agricultural food market, the centre of gravity in which negotiation between supply and demand took place.[10] The major suppliers all lived within the capital or in the immediate outskirts. Indeed, the development of this centralised market was one of the factors which led to the outstanding success of English agriculture in the eighteenth century – one of the most financially successful models in Europe.[11]

The ADM 111 Public Record Office series dated between 1 April 1756 and 31 December 1762 indicates all contracts which were drawn up by the Commissioners of the Victualling Board. The record shows that seventy-one products and/or services were involved (this inventory can be seen in Table 2.2).

The sheer size and variety of this order demonstrates the breadth of activity undertaken by the Victualling Board. Not only was it responsible for general victualling, as can be seen from the orders placed for cereals, meat and beverages; but it was also responsible for the processing of primary products such as cereals into bread and biscuits, oats into porridge, and fresh meat into cured meat; and for the distribution of these goods, which required a whole array of extra materials; and last but not least, short-term contracts had to be drawn up for occasional maintenance duties. The bulk of these supplies were destined for the Victualling Board's London premises, the main centre of victualling because it was the best equipped in terms of food processing (see Chapter 4).

There were basically two types of contract available: one was for periods of time which varied according to need, at irregular intervals, but for specific quantities of foodstuffs; the other type was for regular time-periods (usually periods of three months or one year), but for which the quantities which would eventually be required could not be stipulated in advance. In this case, the suppliers were contractually required to adapt to the fluctuations of demand, and at an agreed price for their goods. Generally speaking, the latter contract seemed to apply mostly to suppliers of fresh meat (fresh meat was bought for seamen who might fall sick and these numbers could not be easily predicted), and to suppliers of butter and cheese – foodstuffs which perished very easily. Last but not least there were contracts for the various services required, and these contracts were remunerated on a *pro rata* basis according to how much work was carried out.

10 London had thirty markets, almost all of which were specialised. Cf. J.-P. Poussou, 'Les villes anglaises ...', in Poussou, *Les Villes en Europe Occidentale* (Paris, CDU, 1983), vol. 2, 108.

11 Cf. F.J. Fisher, *London and the English Economy, 1500–1700* (London, Hambledon Press, 1990).

Table 2.2 Products and services furnished as a result of contracts drawn up by Victualling Board Commissioners

Goods	Regularity
Oats	Irregular
Groats	Irregular
Wheat	Irregular
Wheat for Plymouth	Irregular
Wheat for Portsmouth	Irregular
Wheat for Dover	Irregular
Biscuits	Irregular
Butter and cheese	Annual
Flour	Irregular
Hops	Irregular
Olive oil, Genoa oil	Irregular
Dried amber malt	Irregular
Dried amber malt for Plymouth	Irregular
Peas	Irregular
Stockfish	Irregular
Black Smyrna raisins	Irregular
Rice	Irregular
Fresh beef for Portsmouth	Quarterly
Fresh beef for Jamaica	Permanent Contract
Fresh meat for Thames	Quarterly
Fresh meat for the Downs	Quarterly
Fresh meat for Torbay	Quarterly
Fresh mutton for Chatham and ships at Nore, Sheerness and Blackstakes	
Oxen and hogs for sea store for London	Irregular
Oxen and hogs for sea store for Portsmouth	Irregular
Oxen and hogs for sea store for Plymouth	
Oxen and hogs for sea store for Dover	Irregular
Irish beef	
Irish pork	Irregular
Mutton for the use of the sick and wounded, weather mutton	Quarterly
Malt spirits	Irregular
Sea beer for London	
Sea beer for Portsmouth	Annual
Wine for Lisbon	Permanent Contract
Wine for the supply of HM ships at Oporto	Irregular
Wine for Coruña	Irregular
Wine for Vigo	Irregular
Malaga, Port Vidonia Canary and Fayol wines to be delivered at Guernsey	Irregular
Vidonia Canary wine to be delivered at Tenerife	Irregular
Wine for Portsmouth and Plymouth	Irregular
White salt	Annual
Brine salt made after the late Mr Thomas Lowndes's method and bay salt	Annual

Materials and tools	Regularity
Candles	Annual
Canvas for pudding bags	Irregular
Bulrushes for use in cooperage	Annual
Bulrushes for use in cooperage in Portsmouth	Irregular
Twigs	Irregular
Biscuit bags	Irregular
Casks	Irregular
Buts and puncheons (72 to 120 gallon barrels)	Irregular
Dry hogheads and cheese casks	Irregular
Dry hogheads and dry barrels	Irregular
Hoops per pack of one hundred for six types of cask	
Iron hoops for barrels	Irregular
Wooden hoops per pack of one hundred for six types of cask	Irregular
Coopers tools for London	
Coopers tools for Portsmouth	Irregular
Coopers tools for Plymouth	
Staves for pipe and hoghead barrels	
Staves for big barrels	Irregular
Coals	Irregular
Faggots	Permanent Contract
Wagons to transport cheese to Portsmouth	Irregular
Services	
Sale of the chaff produced by the making of groats and oatmeal	Permanent Contract
Freight	Irregular
Horses and drivers for hulling of oats and grinding groats into oatmeal at Redhouse	Permanent Contract
For taking away oat hulls made at Redhouse	Annual
For taking away the bran milled at Rotherhithe	Annual
For taking away the grains at HM Brewhouse at Hartshorn	Annual
For taking away the small pieces of coal at Redhouse	Permanent Contract
For taking away the sweepings from the various warehouses at Redhouse	Annual

The same procedure in the drawing up of contracts was applied to all sectors, regardless of whether it involved the provision of a given foodstuff, a specific service, or the choice of a contractor for the entire victualling of a secondary home port or an outlying port or station.

The method adopted was as follows: first of all, the Commissioners would announce their invitation to tender bids by posting them at the Royal Exchange and publishing them in a number of newspapers, including the famous *London Gazette*. The pre-publication announcement always took the same form in the minutes of the Victualling Board: *ordered that Monday next,*

5th of January, be appointed to treat for flour. Let it be advertised, publication made and notice given as usual.[12]

There were only two or three days at most between the call for bids and the negotiation of contracts.[13] The only foodstuff that was accorded more time was fresh meat to be furnished at the Thames Estuary – at Chatham, the Nore, Sheerness, Blackstakes and the Downs. In this case, a period of twelve days was permitted so that local traders could get down to London, having been represented, or having sent a letter to the Victualling Board with their proposals.[14]

On the arranged date, the Commissioners met all the traders individually. The traders were asked whether or not they would be able to furnish the quantities required or fulfil an order within a certain time limit. They were asked to produce a sample indicative of the quality to be expected, and to give an indication of their asking price.

Once this first series of offers had taken place, and after the examination of written proposals furnished by traders *in absentia*, the Commissioners then called a second meeting in which the traders were expected to lower their prices. The trader who offered the lowest price won the contract. In the event of several traders offering the same price, several contracts were drawn up and the quantity of goods required was split equally between the traders.

As an example, let us consider the following negotiation taken from the Victualling Board's registers:

> Tenders for wheat for Plymouth, 1000 quarters to be shipped in 14 days. First time: Mr Tho Farrar 75s/per quarter,[15] Mr Jos. Curtis 70s/per quarter; Second time: Mr Tho. Farrar 69s/3d per quarter, Mr Jos. Curtis 68s/11d per quarter. Mr Woods could only offer 300 quarters, so no tender was received from him. The Board didn't think proper to contract for any wheat for Plymouth today the price being so very high, of which the dealers were acquainted.[16]

On this occasion, since the Commissioners had found the going rate for wheat to be too high, a sale was not concluded. However, had it been imperative to buy, the lowest bidder, Mr Joseph Curtis, would have won the contract

12 PRO, ADM 111/51, 2 January 1761.

13 See, for example, PRO, ADM 111/46, 26 June 1758; ADM 111/49, 22 November 1759; ADM 111/51, 2 January 1761. Documents concerning rice, raisins, groats, peas, flour and biscuits.

14 See, for example, PRO, ADM 111/49, 12 and 24 September 1759.

15 The weights and measures which appear in the archive material have been used. Invaluable information on conversion is to be found in Joan Thirsk, *The Agrarian History of England and Wales 1500–1640* (Cambridge, Cambridge University Press, 1967), vol. V:1, 815–25.

16 PRO, ADM 111/ 46, 16 May 1757.

and the Victualling Board secretary would have made an entry in the heavy *quarto* register using the standard formula for acceptance: 'Oath, penalty and all other conditions as usual; let a contract be drawn.' This formula actually referred to the oath that the suppliers had to swear. Unfortunately it has been impossible to find such a detailed contract, since these documents only started to be kept in the ADM 112 archives from 1776. The oaths were drawn up according to a general directive dated 13 November 1747.[17] The contractors had to promise to respect time-limits and the stated quality of their goods, and not to attempt, in any way whatsoever, to corrupt those responsible for the reception of the merchandise: 'Not giving money and as a fee or gratuity to the persons concerned in the receipt of stores.'[18]

The above example shows that Mr Woods, one of the three traders who was unable to supply the requisite 1,000 quarters of corn, but could only provide 300 quarters, did not have his offer retained. This scenario was fairly common and illustrates the difficulty traders had when trying to break into the market. In fact, only fairly well established traders could reasonably hope to obtain contracts.[19] While this approach surely simplified the Victualling Board's task, it also limited the actual number of traders. If a market is transparent enough that it can be well regulated and considered truly competitive, it may not be particularly efficient and the danger is that this situation can give rise to cartels, or at least to oligopolistic behaviour.

High market prices, identical bids, the refusal to come down in price by a shilling or even a penny during the second meeting, were signs that could (and still can) point to complicity between traders. But on only one occasion did the Commissioners specifically refer to this connivance in their records.[20] However, this did not deter them from regularly postponing placing a contract if they deemed the going rate to be too high.

The final section of this book will offer an exhaustive study of the different kinds of contract offered for different kinds of foodstuffs. A breakdown of these contracts season by season can shed light on the extent to which the main merchant companies may have secretly divided up the market sectors. A study of their merger into associations offers complementary information and helps reveal the degree of transparency of the markets.

17 PRO, ADM 111/48, 1 November 1758.

18 PRO, ADM. 111/48, 1 November 1758.

19 Note also PRO, ADM 111/46, 14 September 1757; ADM 111/48, 6 January 1758.

20 PRO, ADM 111/47, 26 June 1758.

Tight State Control

Once a contract had been signed, the Commissioners of the Victualling Board then ensured that the terms of the contract were scrupulously kept to. In those ports in which the Board was directly responsible for victualling, the Commissioners were always informed of any delay in delivery by their Agents. In fact, the Agents were not allowed to accept delivery of a late or incomplete consignment without receiving prior instructions from the Board. In these circumstances there seemed to be basically two options to be taken which were dependent on the going price and on how well stocked the reserve stores were. Thus, the Commissioners could either grant the trader an extension so that he could fulfil his contractual obligations, or they could order their staff to buy the victuals and charge the cost to the contractor. This second option was particularly distressing, financially speaking, for the latter.

A single example will suffice to illustrate the determination with which the Commissioners expected their contracts to be honoured. On 5 August 1757, Messrs Samuel and Isaac Solly signed a contract to supply 2,000 quarters of wheat which were to be delivered to London in three deliveries before the end of the month. The price agreed upon was 31 shillings per quarter.[21] Since the traders had not furnished the agreed quantity by 17 August, part of the outstanding amount was bought from another trader, a Mr Farrar, at the price of 37 shillings and 6 pence per quarter.[22] By 26 August, when the Sollys' second delivery had also not materialised, a further 300 quarters of wheat were again purchased from Mr Farrar, but this time at the rate of 46 shillings per quarter. The Solly brothers were ordered to pay the difference between their original contract price and Mr Farrar's price, i.e. 8 shillings and 6 pence.[23]

In the meantime, Samuel and Isaac Solly had apprised the Commissioners of their difficulty in meeting their obligations, and had asked to be released from them,[24] to which the Victualling Board replied by mounting an immediate attack. The wholesalers were summoned and discovered that it was out of the question to be released from a contract that they had freely entered into. The Commissioners then ordered them to deliver the final third of the order by the 27th of that month.[25] When, by 2 September the wheat still had not been delivered, it was once again purchased from Mr Farrar, but this

21 PRO, ADM 111/46, 5 August 1757.

22 PRO, ADM 111/46, 17 August 1757.

23 PRO, ADM 111/46, 26 August 1757.

24 PRO, ADM 111/46, 26 August 1757.

25 PRO, ADM 111/46, 26 August 1757.

time at the price of 54 shillings per quarter.[26] This episode must have proved a salutary lesson for future traders not to engage rashly.

Furthermore, all foodstuffs supplied to the principal ports under the Victualling Board's jurisdiction were controlled not only in terms of quantity but also of quality. The victuals were delivered in containers and according to well-defined regulations, aimed at simplifying the Victualling Board's accounts. A certain number of the sacks and casks were opened to verify that the contents were of the required quality. If and when this was found not to be the case, the whole consignment would be opened, and a report would then be filed and sent to the Commissioners. If a clause existed in the contract which specified a minimum length of time in which the victuals should remain fresh, the Commissioners would then deduct the value of the sub-standard or unacceptable victuals from any outstanding bills they may have had with the trader. Correspondence between the Commissioners and their Agents reveals that this verification procedure was not taken lightly.

On only one occasion, in all the war and over the totality of products supplied to the Board's own yards, the Commissioners came close to terminating an agreement on the grounds of breach of contract, which gave the Board the legal right to bring proceedings against the contractor in order to obtain damages, and bring the agreement to a definitive end. The incident occurred at the beginning of 1761 and concerned Messrs Barnevelt and Townshend, sole suppliers of butter and cheese to all the yards directly under the jurisdiction of the Victualling Board. Pursuant to complaints lodged by Board Agents in Portsmouth and Plymouth regarding the rapidity with which the cheese became rancid,[27] the Commissioners ordered (on 28 January 1761) that an investigation should be promptly carried out in each of the principal ports (London, Portsmouth, Plymouth, Dover and Chatham). The following instructions were sent to the Board's Agents in each of these ports, and underline the rigour with which this quality control was to be carried out:

> In conjunction with our officers to examine all the cheese in the stores and they find any part thereof not be answerable to the terms of the said contract they to lay the same apart and send us up an account of the quantity thereof and to take notice of the marks and numbers of the casks out of which the said consommation shall have been taken, and also when and by what vessels the said casks were sent from hence.[28]

26 PRO, ADM 111/46, 2 September 1757.

27 PRO, ADM 111/51, 26 January 1761.

28 PRO, ADM 111/51, 28 January 1761.

On 3 April, the contractors were asked to appear before the Commissioners of the Victualling Board,[29] and on 27 April they were informed of the damning results of the inquiry. In each cask there had been found some small, rancid cheeses, packed in such a way as to leave no doubt that their inclusion had not been accidental. The Commissioners thus informed Messrs Barnevelt and Townshend of their decision to sue for breach of contract.[30]

The wholesalers then insisted that the cheeses had been placed in the casks by their employees *who had not told them about it* but who had acted in good faith. In fact, it was common practice, they explained, to place small rancid cheeses in the middle of casks in order to fill up the gaps. They ended their defence by declaring that they were willing to shoulder all the financial consequences of the unfortunate affair, and begged the Commissioners not to take the case to court, pleading that the situation would never arise again. Reproduced below is an extract from their deposition:

> Upon which Mr. Barnevelt said that those cheese had as he informed them been put up by his servant without his knowledge and that as to there being only 2 or 3 in a cask, he alledged that when they packed these cheese with cheshire, they are obliged to put the thin cheese in the middle of the cask, on account of their widen there, than at the ends. That as to the rest, his partner and him were very ready to take the cheese back and pay for it and that as the quantity of it was by small, he hoped the Board in regard thereto, and to the affair having happened in a hurry and through the inadvertence and mistake of their servant would please not to order any proceedings at Law upon it, but to permit the cheese to be returned to them which they would repay the cost of, and would take particular care that nothing of that kind should happen again in the future.[31]

We do not know whether or not Messrs Barnevelt and Townshend were eventually prosecuted. However, it seems unlikely that they were able to tender for new contracts.[32]

As far as secondary home ports and outlying ports and stations were concerned, in which the Victualling Board did not have any Agents but relied on a single supplier for all their victualling needs, the Commissioners could not assure any form of quality control. They had to rely on communications from squadron commanders or other officers informing them of

29 PRO, ADM 111/51, 3 April 1761.

30 PRO, ADM 111/51, 27 April 1761.

31 PRO, ADM. 111/51, 27 April 1761.

32 Messrs Smith and Copeland obtained the last two annual contracts during the Seven Years War. See PRO, ADM 111/52, 24 August 1761; ADM 111/54, 25 August 1762. In 1758 the private contractor for Nova Scotia was taken to court for breach of contract (see Chapter 6 below).

any missing or inadequate products. The wholesaler would then be contacted and informed of the negligent manner in which he was doing business. On 9 March 1762, for example, Captain Forrest stationed in Jamaica complained about the shortage of bread. The suppliers – a company called Nesbitt and Franks – were thus immediately requested to make available to the Commissioners a breakdown of their supplies to this station, and their proposal on how to remedy the shortage.[33] Four days later, on 13 March, Messrs Nesbitt and Franks replied to every point raised by the Board. In addition, they had instructed their assistant in New York to send 'with all possible despatch' a load of bread to Jamaica, on top of the quantity sent every six weeks from New York.[34] By 22 March, the mystery surrounding the missing bread was revealed: in fact, the supplier's integrity was not in question at all – quite simply, one of his ships had been captured. In order to avoid a repetition of the bread shortage, the Commissioners ordered Nesbitt and Franks to reduce the intervals between their delivery cycles from New York. From then on, they had to take place every month and 'carry a greater proportion of bread and other victuals necessary for the good victualling of 6000 men in Jamaica'.[35]

In cases of shortage, the captains of ships in secondary home ports and at outlying ports and stations were naturally expected, just like the Board's Agents in the principal ports, to procure the necessary victuals locally by a bill of exchange, which amount was deducted by the Navy from sums due to the contractor concerned.

The three main victualling centres – London, Portsmouth and Plymouth – were under the jurisdiction of the Victualling Board and were equipped with the necessary infrastructure to allow the Board to control its own food treatment and processing, as a consequence permitting it to better manage the food chain. The Commissioners were unceasing in their application of the correct procedures and regulations.[36] Most of the Navy's victuals passed through London, and in order to render the inspection process more efficient, a new post was created just before the onset of hostilities, that of Victualling Officer. Consequently, a Mr John Kelsall found himself based in Deptford, where most of the victuals were processed. Kelsall, an experienced 59-year-old official, began his duties as Victualling Officer on 13 March 1756[37] and was responsible for helping the main Agent at the port to supervise all its victual-

33 PRO, ADM 111/53, 9 March 1762.

34 PRO, ADM 111/53, 13 March 1762.

35 PRO, ADM 111/53, 22 March 1762.

36 See, for example, the investigation following a complaint lodged against Allen Spencer concerning a shipment of beef: PRO, ADM 111/51, 27 February, 23 March, 20, 22 and 24 April 1761.

37 PRO, ADM 110/20, 3 March 1761.

ling activities and to ensure that the Board's suppliers and its own employees carried out their duties scrupulously.

The Board did not confine itself to simple admonitions. When circumstances demanded, it did not hesitate from dismissal. Such was the case for Richard Herring and Gabriel Small, master brewers at Plymouth and Deptford who were dismissed from their posts in September[38] and November[39] 1756 respectively. In February of the same year, Henry Shales, the Victualling Board's Agent in Portsmouth – that is to say, the principal victualling officer in the port – was also dismissed following a number of complaints lodged by Admiral Boscawen.[40]

In addition to checking and inspecting shipments, the Victualling Board was also responsible for the expediting of victuals that had been negotiated in London and delivered to Deptford, to the other ports under its jurisdiction, namely Portsmouth, Plymouth, Dover and Chatham. The distribution of consignments to these ports did not necessarily occur at regular intervals, but instead consignments were dispatched at the request of the Victualling Board Agent stationed in the port in question, whenever he saw the need. Table 2.3 groups together the transfers ordered by the Commissioners between 16 May and 31 December 1757, with a breakdown by destination, product and quantity.

If any victuals in one of the ports were in danger of spoiling due to the length of time they had been stocked, then the Agent in that port, having obtained permission from the Board, could sell them. The money obtained from these transactions had to be submitted to the cashier of the Victualling Board.[41] This leads to consideration of the financial aspects relating to the management of the Victualling Board during the Seven Years War.

Financial Practices

The Victualling Board employed at least twenty clerks who were responsible for accounting and finance.[42] Their managerial superior was the *cashier for the affairs of the victualling* to whom all questions and queries were directed. All Treasury Board monies ended up in his service, having arrived via the Navy Pay Office. Directed by the Treasurer of the Navy, the Navy Pay Office was

38 PRO, ADM 110/28, 6 September 1756.

39 PRO, ADM 110/19, 11 November and 5 December 1757.

40 PRO, ADM 1/294, Boscawen, dated 9 February 1757. He was replaced on 14 February by Matthew Oakes, aged 41 (PRO, ADM 110/20, 3 March 1761).

41 PRO, ADM 111/46, 23 May 1757.

42 PRO, ADM 20/225, 31 March 1757. See also Appendix 1.

Table 2.3 Produce sent from London to Portsmouth, Plymouth, Chatham and Dover between 15 May 1757 and 31 December 1757

Product	Unit	Portsmouth	Plymouth	Chatham	Dover	Total
Cheese	pound	367,000	200,000	27,000	30,000	624,000
Butter	pound	140,000	80,000	12,000	13,000	245,000
Olive oil	gallon	6,000	2,000	/	/	8,000
Vinegar	gallon	22,000	16,000	500	/	38,500
Beef	piece	150,000	50,000	6,200	/	206,200
Pork	piece	225,000	100,000	6,300	/	331,300
Peas	bushel	13,000	8,000	900	800	22,700
Groats	bushel	5,000	2,100	/	500	7,600
Hops	bag	370	280	/	/	650
Raisins	pound	65,000	10,000	/	/	75,000
Bread	bag	14,000	22,000	1,200	/	37,200
Suet	pound	8,000	10,000	/	/	18,000
Beer	tun	/	/	200	/	200
Brandy	gallon	10,000	/	/	/	10,000
Flour	pound	/	234,000	5,000	28,000	267,000
Oatmeal	bushel	/	1,000	/	/	1,000
Delivery dates		6, 15, 29 June; 13 July; 5, 10, 24 Aug; 9, 19, 21 Sept; 5, 7 Oct; 2, 18, 23 Nov; 9, 19, 21, 23 Dec	6, 15, 29 June; 13, 25 July; 10, 24 Aug; 9, 19, 21 Sept; 7 Oct; 2, 11, 23 Nov; 9, 23 Dec	6, 15 June; 25 July; 10 Aug; 21 Sept; 17 Oct; 2, 23 Nov; 23 Dec	29 June; 10, 24 Aug; 5 Oct; 21, 23 Nov	

Source: Table drawn up from data in ADM 111/46.

theoretically hierarchically subordinate to the Lords of the Admiralty (see Appendix 3), but in practice it was becoming more and more independent and issued funds – whenever it could, and after strict verification – to the different Boards (the Victualling Board, the Navy Board, the Sick and Wounded Board, etc.).[43]

It was thus up to the cashier of the Victualling Board to issue payments[44] and to ensure that the Treasurer of the Navy received all the receipts. These were then assembled by the Pay Office and copied into ledgers on the 14th and last day of each month.[45] These statements, which were closed every few years, allowed the Victualling Board's books to be balanced every calendar year.

Some abstracts have survived and summarise the receipts and expenses of each of the large yards. The receipts covered five main areas: fees received for 'old stores sold and rent received' by Victualling Board Agents; arrears payments 'for debts on victualling accounts'; deductions for non-performance of contractual obligations – 'abatements from bills'; deductions for provisions found spoilt within the guarantee period – 'provisions cast by survey within time of warranty'; and finally, the refunding of advances or 'imprests cleared'. The expenses were subdivided into thirty-seven budget items and these are listed with their annual costs in Appendix 4.

For example, the sum of £1,053,163 in 1759 (which represented 20.73 per cent of the Navy's total expenditure[46]) consisted of meat which was bought for the ports of London, Portsmouth, Dover and Plymouth and came to £265,570 (25.22 per cent), cereals and pulses worth £163,620 (15.54 per cent), £35,847 (3.4 per cent) for casks and hoops, as well as cartage and carriage, freight and demurrage worth £21,671 (2.06 per cent), and salary and wages worth £70,261 (6.67 per cent). Finally, the victualling of the secondary home ports and the outlying ports and stations, for which the Victualling Board relied on private contractors, represented 19.46 per cent of total expenses in 1759, that is to say £204,940.[47]

43 Gradish, *The Manning of the British Navy*, and N.A.M. Rodger, *The Admiralty* (Lavenham, Suffolk, Terence Dalton, 1979).

44 As a rule, the Victualling Board rounded up when payments were made to its own personnel, and rounded down when suppliers were paid.

45 PRO, ADM 20. The payments ledger was of particular importance at the end of every third month, due to the payment of personnel.

46 £5,080,263 was spent by the Navy in 1759. See PRO, ADM 49/47: a table of expenses incurred by the Royal Navy between 1755 and 1763.

47 PRO, ADM 20/235. When a Royal Navy vessel dropped anchor at a place where no means of victualling had been set up, the purser could, when necessary, go through a third party to obtain the missing foodstuffs. Throughout the duration of the war, these

Figure 2.1 Number of days beyond six months non-payment during which the Victualling Board incurred interest charges

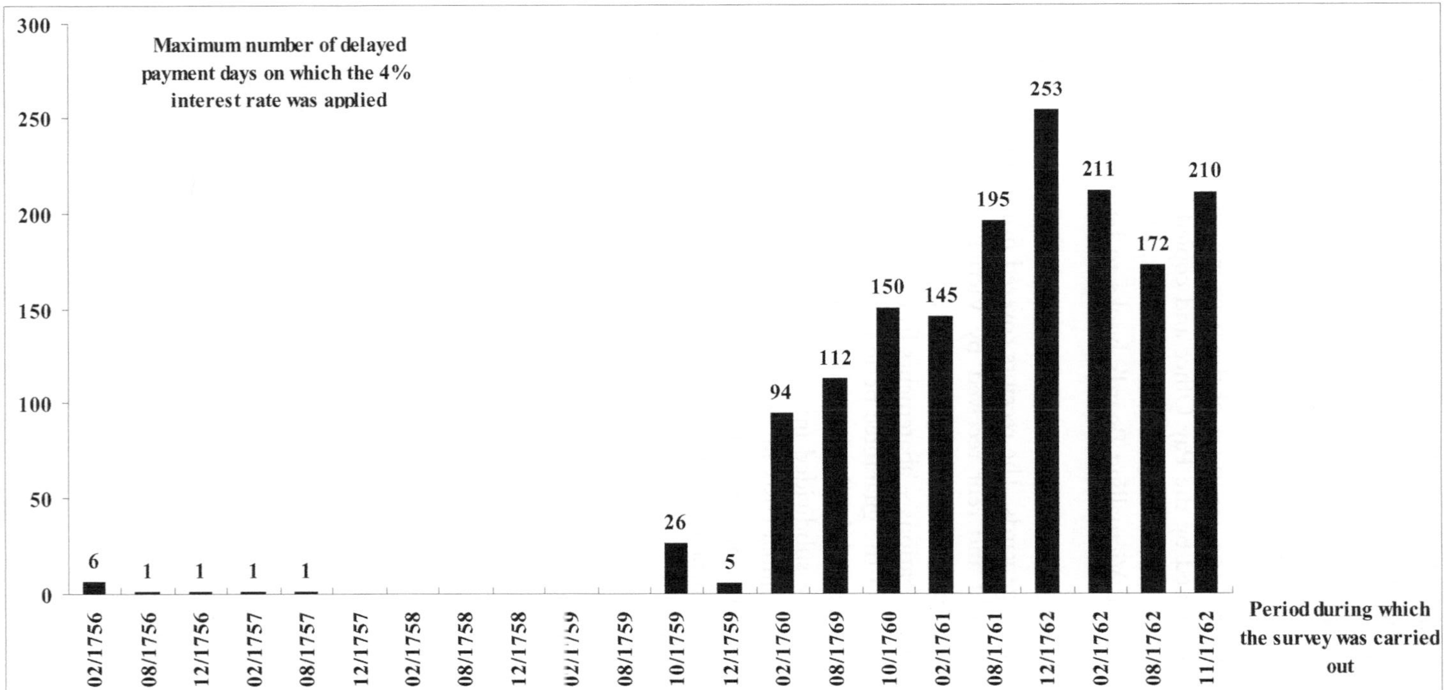

The ADM 20 ledger series also permits an evaluation of the relative difficulty the Crown had in financing this expenditure, and allows an examination of England's financial capacity to follow through with, or even intensify, the war effort. However, it is first necessary to describe how payment took place and how it evolved.

Since the Victualling Board was (generally) unable to pay its suppliers' bills as soon as they arrived, it had developed (in the 1680s) and maintained a very fair system in which the invoices were numbered as they arrived and paid in that order. Outstanding debts of more than six months were subject to an interest charge of 4 per cent, applicable between the years 1749 and 1783. In fact, this interest rate varied: it was at 6 per cent between 1683 and 1713, at 5 per cent between 1714 and 1748, and then at 5 per cent again from 1784 to 1795.[48] If a supplier could not wait to be paid, it was up to him to contact one of the many London brokers who would exchange his letter of credit, minus a certain discount, for the sum owed to him by the Navy. These discount rates were applied specifically to Victualling Board contracts and were periodically published in the newspapers in 1750, and regularly from 1758.

Clearly, the prices proposed by merchants when the invitations to tender bids were published took into account this commonly known delay in being paid, and also absorbed the discount rate. Any further delays in payment necessarily resulted in an increase in the discount rate which in turn inflated the prices proposed to the Victualling Board. Figure 2.1, which covers the seven years of the war, shows the number of days on which non-payment by the Victualling Board (after the allowed six-month period) was subject to a discount rate surpassing 4 per cent. The figure also shows the difficulties encountered in terms of the Navy's financing.

It should be noticed that in spite of the rapid rise in demand following the outbreak of hostilities, 1756 and 1757 were years in which almost all payments were met in full within the allotted six months following reception of the invoice. In fact, there were only a few instances of late payment: two of six days in February 1756,[49] and a few others of one day in August and December 1756, and February and August 1757.[50] Surveys carried out on the December 1757 to August 1759 period reveal that late payment stopped altogether, which illustrates the extent to which the state was capable of financially supporting the Navy. However, from October 1759, fewer and fewer invoices

middle-men received a commission worth 2.5 per cent of the total value of victuals bought. See, for example, ADM 20/228, invoice 1295; ADM 20/246, invoices 657 and 672.

48 Sir W. Beveridge, *Prices and Wages in England from the Twelfth to the Nineteenth Century* (London, Frank Cass, 1965), 519.

49 PRO, ADM 20/225, invoice nos. 185 and 198.

50 PRO, ADM 20/225 and 228.

were being settled in the set six-month period. By February 1760, most invoices were being settled with compensatory interest payments for 94 days. In 1760 and 1761 some invoice settlements were being delayed by as many as 253 days over and above the allowed six-month period. However, 1762 was a slightly better year, with delays of 211, 172 and 210 days in February, August and November.[51] Payment was thus taking place with delays of over a year, the first six months of which were interest-free.

This development was mirrored by discount rates which, from 2.5 per cent in January 1760, reached a high of 12.5 per cent in February 1762 (see Table 2.4). These successive increases in interest rates also led to an increase in the prices of the different foodstuffs. This is an important point to be taken into consideration when examining price fluctuations per product.

Table 2.4 Fluctuations of interest rates as applied by Victualling Board creditors

	1758	1759	1760	1761	1762
January	2⅛–2¼	2½–2⅞	4⅞–5¾	9–10	10¾–12¼
February	1½–2	2⅝–2⅞	4¼–7½	9¼–9⅞	10¾–12½
March	1⅜–1⅝	2½–2⅝	6⅝–8	9⅞–10½	8¼–10½
April	1⅝–1¾	2⅝–4½	8–9	7–10¾	8¼–10
May	1⅝–1⅞	3½–4¼	8¾–9	6–7¼	9¾–10
June	1¾–1⅞	3–3¾	9½–10	6⅞–7¼	9¼–9¾
July	2–2⅛	2⅞–3½	8⅜–9¾	7⅜–8	8¾–9½
August	2⅜–2½	3¾–4	6½–8⅝	8½–10⅛	7¾–8½
September	2¼–2⅜	3–4⅝	6⅝–7⅞	8¾–9¼	4¾–6½
October	2¼–2½	3¾–5¼	6½–8¾	9¾–10⅝	6¾–8⅜
November	2½	4¾–5½	7¾–9	9⅝–9⅞	5½–8⅛
December	2½–2¾	4¾–5	8⅜–9⅜	9⅜–10¾	3⅞–5¾

Nonetheless, the price lists published by Sir William Beveridge should be considered with great caution; especially since his only sources for the period under consideration were the ADM 20 ledgers. In fact, the invoices found in these ledgers do not give the dates of the contracts which means that the dates on which the contracts were made are unknown. All that is known is the delivery dates, the date on which the invoice was registered and the date on which it was paid. As a consequence, the recorded prices do not correspond exactly to the periods given in Table 2.4: those prices are in fact the prices at delivery and not the set prices. This difference makes a precise study of prices all the more difficult as the variation is not constant over time and

51 Figures taken from ADM 20/225, 228, 231, 235, 238, 242, 244, 246 and 248.

fluctuates according to the particular foodstuff. The third part of this book will therefore offer an analysis of the markets and their rates, using the ADM 111 series which records the dates on which contracts were signed and the market rates on those days. This analysis should be seen as complementary to the immense work carried out by Lord Beveridge, and should in no way be considered as putting it into question. Indeed, Beveridge's publication of prices over the time-span of a century has equipped later historians with the tools with which to compare the various areas of the economy at a particular point of time and over a period of years.

3

Food Rations and Their Evolution

When British maritime historiography refers to the issues with which victualling was confronted before the 1790s, it has usually painted a dark and often erroneous picture. The same quotation is nearly always cited, from William Thompson's 1761 work entitled *An Appeal to the Public in Vindication of Truth and Matters of Fact*. The author, who was a foreman cooper for the Victualling Board in the 1740s, vividly describes the deplorable condition of foodstuffs:

> That Mariners in the King's Ships have frequently put their 24 hours' allowance of salt provisions into their tobacco-boxes. That seamen in the King's Ships have made buttons for their Jacketts and Trowses [*sic*] with the cheese they were served with, having preferred it, by reason of its tough and durable quality, to buttons made of common metal; and that Carpenters in the Navy-Service have made Trucks to their Ships' flagstaffs with whole Cheeses, which have stood the weather equally with any timber. That the Flour in the King's Ships has been devoured by weevils, and become so intolerably musty, and cemented into such hard rocks, that the men have been obliged to use instruments, with all their feeble power, to break and pulverise it before they could make use of it, as though, in a comparative degree, they had been stubbing to pieces the ruins of an old fortification.
>
> That their bread has been so full of large black-headed maggots and that they have so nauseated the thoughts of it, as to be obliged to shut their eyes to confine that sense from being offended before they could bring their minds to a resolution of consuming it.
>
> That their beer has stunk as abominably as the foul stagnant water which is pumped out of many cellars in London at midnight hour; and that they were under a necessity of shutting their eyes, and stopping their breath by holding of their noses before they could conquer their aversion, so as to prevail upon themselves in their extreme necessities to drink it.[1]

These remarks quickly brought discredit to the Victualling Board Commissioners who were judged at best as being incompetent and negligent, at worst as deceitful and corrupt. They countered by providing simplistic and convenient explanations for the high mortality and morbidity rates that continued

1 H.W. Hodges and E.A. Hughes, eds, *Select Naval Documents* (Cambridge, Cambridge University Press, 1922), 125–6.

to ravage the Navy until the end of the century. Should they also be held accountable for the late introduction of lemons in daily rations, which, thanks to their rich source of vitamin C, began to play an important role in the fight against scurvy in the 1790s?

It should not be forgotten that William Thompson's remarks were of an extreme nature, and that he was dismissed from the Victualling Board in January 1745 following the discovery of spoiled and rotting provisions that he had been in charge of.[2] His allegations therefore need to be treated with circumspection. Credit should be given to Nicholas Rodger for his article in which he offers a study of Thompson's character and written output.[3] The partisan and political nature of Thompson's writings is made clear in his choice to dedicate his *Royal Navy Men's Advocate* (1757) to Alderman Beckford, the leader of the opposition to the Newcastle administration.

Professor Daniel Baugh, in his notable study of naval administration in Walpole's time, praises the work of the Victualling Board up to the end of the War of the Austrian Succession.[4] Indeed, study of the food rations issued during the Seven Years War in this chapter's first section gives rise to a re-evaluation of past criticism. They now appear to have been quantitatively sufficient and qualitatively sound, and the extent of the innovations in terms of food management in this period invites reconsideration, in the second section, of the commonly held date at which the food revolution in maritime matters began. The third section uses the example of the western squadron to illustrate the close, interactive links that existed between victualling and strategy.

Quantitatively Sufficient and Qualitatively Sound Food Rations.

In 1677, the weekly menu for seamen serving with the Royal Navy was drawn up and then appears to have changed very little. It was only with the founding of the Victualling Board on 4 January 1701 that some limited variation occurred. Until then, pork had simply been a substitute for beef when supplies had run out. Now, pork was included in normal rations and was served twice a week, as was beef.[5] Fish, on the other hand, which was very

2 W. Thompson, *The Royal Navy Man's Advocate* (London, 1757); W. Thompson, *The Case of William Thompson* (London, 1758).

3 N.A.M. Rodger, 'The victualling of the British Navy in the Seven Years' War', in *Bulletin du Centre d'Histoire des Espaces Atlantiques* (new series no. 2, Talence, Maison des Sciences de l'Homme d'Aquitaine, 1985), 37–54.

4 Baugh, *British Naval Administration*. See also *Naval Administration 1715–1750*.

5 Merriman, *Queen Anne's Navy*, extract no. 1, 254–8.

expensive,[6] disappeared from the weekly menu and became a substitute meal if the meat rations ran out. Fish was replaced by oatmeal.[7]

This menu, which was to last for most of the eighteenth century, is summarised in Table 3.1.

Table 3.1 Weekly rations

	Biscuits (lbs)	Beer (gallons)	Salt beef (lbs)	Salt pork (lbs)	Peas (pints)	Porridge (pints)	Butter (oz.)	Cheese (oz.)
Sunday	1	1	-	1	½	-	-	-
Monday	1	1	-	-	-	1	2	4
Tuesday	1	1	2	-	-	-	-	-
Wednesday	1	1	-	-	½	1	2	4
Thursday	1	1	-	1	½	-	-	-
Friday	1	1	-	-	½	1	2	4
Saturday	1	1	2	-	-	-	-	-

Source: PRO, ADM 110/22, p. 316.

The nutritive value of these rations does deserve to be underlined, even though they were far from being well balanced, as attested by the almost total lack of vitamin C. J.C. Drummond and Anne Wilbraham[8] and, subsequently, Christopher Lloyd[9] agree that this menu provided a daily quota of 4,000 to 4,500 calories. Taking into account the fact that pursers were authorised to reduce rations by an eighth to allow for wastage, it would still appear that the rations of naval crews were more than sufficient and indeed superior to the diet of an ordinary land-dweller.[10] For example, the opportunity to eat meat four times a week was rare at the time,[11] and contributed to making the naval service an attractive career option.[12] Last but not least, another very popular measure was the allocation of beer, in theory limited to one gallon per day per man; in reality unlimited.[13]

6 Baugh, *British Naval Administration*, 378.

7 Merriman, *Queen Anne's Navy*, extract no. 1, 254–8.

8 J.C. Drummond and A. Wilbraham, *The Englishman's Food* (London, Jonathan Cape, 1939).

9 C. Lloyd and J.L.S. Coulter, *Medicine and the Navy* (Edinburgh, Livingstone, 1961); C. Lloyd, 'Victualling of the fleet in the 18th and 19th centuries', in *Starving Sailors*, ed. J. Watt, E.J. Freeman and W.F. Bynum (London, National Maritime Museum, 1981).

10 Cf. G.E. Fussel, 'The change in farm labourer's diet during two centuries', *Economic History Review*, I (1927), 269–74.

11 T.S. Ashton, 'Changes in standards of comfort in 18th century England', *Proceedings of the British Academy*, XLI (1995).

12 In the opinion of Rodger (see *The Wooden World*, 87).

13 Baugh, *British Naval Administration*, 376–7.

On the basis of these weekly rations (see Table 3.1), a mathematical formula was applied to calculate the quantities of food required for each vessel about to set sail, based on the number of men on board and the length of the campaign. The register of shipping in ADM 30/44 provides a very interesting set of documents relating to the management of victuals, which has allowed scholars to reconstruct the logic behind the calculations. For example, in order to arrive at the number of pounds of bread and gallons of beer required, it sufficed to multiply the number of seamen by the estimated length of the crossing. If the result was then divided by seven, the number obtained indicated the required quantity of salt beef and pork; by dividing that number by four, the figure obtained represented the number of gallons of peas that were needed. If 50 per cent was added to that number, the sum obtained represented how many gallons of oatmeal and pounds of butter were needed. Last but not least, multiplying the last obtained figure by two resulted in the number of pounds of Suffolk cheese that were needed. Table 3.2 summarises the above calculations.

Table 3.2 Calculation of the quantities of victuals required

Changeable Parameters		
Number of men	100	
Estimated length of the campaign	28 days	
Victuals		
Pounds of bread and gallons of beer	100 x 28	= 2,800
Number of pieces of salt beef and pork (4lb pieces of beef, 2lb pieces of pork)	2,800 ÷ 7	= 400
Peas (gallons)	400 ÷ 4	= 100
Porridge (gallons) and butter (pounds)	100 + 50	= 150
Suffolk cheese (pounds)	150 x 2	= 300

A second type of rations was also planned in parallel with the above type, this time adapted to long ocean crossings. It paid special attention to foodstuffs with a high calorie content, for these were known to perish less easily and to better endure the heat. This particular diet became standard fare on all ships which – according to the terms of the regulations – crossed 'south of the 39th degree of latitude',[14] that is to say, bound for Africa, America or Asia. The Admiralty provided the Victualling Board Commissioners with a series of guidelines on 4 January 1701. In these guidelines, it was stated that

14 Instructions from the Admiralty to the Navy Board (4 January 1701), reproduced in Merriman, *Queen Anne's Navy*, extract no. 1, 254–8.

half the quantity of beer should be replaced by brandy (a half pint of brandy for every gallon of beer), or, in exceptional circumstances and only if brandy was unavailable, by 'beverage wine' (local wine mixed with water). The latter was judged to be less beneficial to good health than beer. In the West Indies where 'brandy is a rare commodity and it has become habitual to distribute three-quarters of a pint daily to the men, plus a pound of sugar', the Admiralty upheld this local practice.

A piece of beef or pork or a portion of bread could be replaced by 3 pounds of flour, or by one pound of grapes or one and a half pounds of currants or one and a half pounds of pickled beef-suet. Four pounds of Milan rice or two fish – which had to be at least 16 inches long – could replace a portion of cod – which had to be 24 inches long. A pint of olive oil could replace one pound of butter or two pounds of Suffolk cheese. Biscuits could be replaced by rusks.[15]

The Admiralty also decreed in July 1740 that half a pint of vinegar per week per man should be allocated on long-haul trips since vinegar was considered to be an antiscorbutic.[16] From 1754, permission was granted to replace oil or butter with sugar on long voyages.[17]

In this way, a table of equivalences was drawn up in a most rational manner. The acceptance of such interchangeability in foodstuffs meant that the Victualling Board was confronted with a much wider range of victuals to procure. However, this, in turn, also allowed for more flexibility, so that the Commissioners could more easily handle certain foodstuff shortages by replacing them with acceptable alternatives, and they were also in a better position to negotiate prices. The flexibility resulting from the possibility of replacing certain victuals with previously established alternatives certainly played a major role in the successful victualling of the British naval forces during the Seven Years War. Table 3.3 shows official foodstuff alternatives.[18] Crews making long voyages also benefited from immediate provision of fresh meat on their return.[19]

Another important measure to take into account in the victualling of vessels sailing to distant waters was the practice of *short allowance*. This consisted of economising on food supplies by reducing each component of a crewman's

15 Admiralty Board recommendations (4 January 1701), reprinted in Merriman, *Queen Anne's Navy*, extract no. 1, 254–8.

16 Admiralty order of 3 July (cf. ADM 110/12, 20 February 1741 and ADM 110/14, 13 July 1744).

17 Beveridge, *Prices and Wages*, 528.

18 PRO, ADM 30/44, p. 15.

19 Admiralty instructions to the Navy Board (4 January 1701), reprinted in Merriman, *Queen Anne's Navy*, extract no. 1, 254–8.

Table 3.3 Foodstuff alternatives

1 gallon beer	= 1 pint wine or ½ pint brandy or rum
1 lb butter	= 1 pint oil
1 piece beef	= 4 lbs raisins or 2 lbs pickled beef-suet
1 piece beef	= 3 lbs flour + ½ lb beef-suet
1 piece beef	= 3 lbs flour + ¼ lb beef-suet + ½ lb raisins
1 pint oatmeal	= 1 lb pearl barley
3 pints oatmeal	= 2 pints peas
3 pints oatmeal	= 1 quarter calavances (haricot beans)
1 gallon oatmeal	= 4 lbs fish or 1 gallon wheat or 4 lbs rice
1 pint oil	= 1 lb sugar
1 lb bread	= 2 lbs potatoes or 2 lbs yams
1 lb butter	= 1½ lbs sugar or ½ lb cocoa
3 lbs Suffolk cheese	= 2 lbs Cheshire cheese

rations by a third. In this way, six men could be fed with the usual quantity of food for four. In lieu, the men received financial compensation. Since the Victualling Board claimed to insist upon the prompt payment of this compensation to the men, it allowed the men, when in port, to buy for themselves the local fresh produce that their ships so desperately lacked.[20]

It goes without saying that the quantities and contents of victuals on board vessels on long-distance voyages differed. However, clearly defined ratios which were established according to destination became the norm. Table 3.4 shows the types of foodstuffs and quantities required on a third-rate vessel carrying 500 men.[21]

Having examined the sea victuals of crews sailing in the European seas, and those of vessels headed for far-distant zones, the third and last type of rations to be considered are those known as *petty warrant*, destined for seamen whose ships lay berthed at the principal naval bases. This regimen differed on only two points from that of ships in European waters. Bread was supplied instead of biscuits, and on a twice-weekly basis the men were able to consume fresh beef, from a recently slaughtered animal,[22] rather than salted meat. However, it appears that this recommendation was not systematically respected. The Commissioners themselves were aware of the fact that if a ship was returning

20 Victualling Board Commissioners to the House of Commons (December 1703), reprinted in Merriman, *Queen Anne's Navy*, extract no. 11, 269–70.

21 See also Appendix 5 which reproduces the list of victuals loaded onto the *Southsea Castle* bound for the Indies in 1759.

22 Victualling Board Commissioners to the House of Commons (December 1703), reprinted in Merriman, *Queen Anne's Navy*, extract no. 11, 269–70.

Table 3.4 Recommended produce and quantities according to distance covered by a third-rate vessel with 500 men aboard

Destination	Bread (pounds)	Alcohol (gallons)	Beer (tuns)	Salt beef (pieces)	Salt pork (pieces)	Flour (pounds)	Suet (pounds)	Raisins (pounds)	Peas (bushels)	Porridge (bushels)	Olive oil (gallons)	Vinegar (gallons)
East Indies	8 months or 112,000 lbs	2 months or 1,750 gal	6 weeks or 87.5 t	6 months or 12,000 p	8 months or 16,000 p	2 months or 12,000 lbs	10 months or 10,000 lbs	/	8 months or 500 bl	5 months or 187.5 bl	3 months or 562.5 gal	12 months or 1,500 gal
West Indies	4 months or 56,000 lbs	10 weeks if orders given or 2,187.5 gal	as above	2 months or 4,000 p	4 months or 8,000 p	as above	2 months or 1,000 lbs	2 months or 2,000 lbs	4 months or 250 bl	4 months or 375 bl	1 month or 187.5 gal	4 months or 500 gal
North America	as above	as above	6 weeks if orders given or 87.5 t	4 months or 800 p	6 months or 1,200 p	as above	as above	as above	as above	as above	as above	as above
Channel	3 months or 42,000 lbs	/	as above	6 weeks or 3,000 p	3 months or 6,000 p	6 weeks or 9,000 lbs	6 weeks or 750 lbs	6 weeks or 1,500 lbs	3 months or 187.5 bl	3 months or 281¼ bl	/	3 months or 375 gal
Mediterranean	4 months or 56,000 lbs	10 weeks if ordered or 2,187.5 gal	as above	5 months or 10,000 p	4 months or 8,000 p	2 months or 12,000 lbs	2 months or 1,000 lbs	2 months or 2,000 lbs	4 months or 250 bl	4 months or 375 bl	/	4 months or 500 gal

Source: ADM 30/44.

to port with salt beef still in stock, it would continue to be served rather than fresh meat.[23]

The practice of *short allowance*, in the manner of that on long-haul voyages, was frequently utilised to give those who desired it the opportunity to buy their own 'refreshments'.[24] However, it is difficult to know whether or not the men actually took advantage of this system.[25]

The above-mentioned rations were destined purely for seamen. Officers, on the other hand, were totally free to purchase their own food. They would either buy their favourite fare themselves or via their servants, or ask the ship's purser to make the purchase and deduct the sum from their pay. The only contribution the Admiralty made to their diet was to provide each officer, according to his rank, with a certain number of ox tongues, considered to be amongst the finest cuts of meat. Table 3.5 shows the number of ox tongues allocated per officer according to rank.

Table 3.5 Number of ox tongues per senior officer

To an Admiral of the Fleet	7 dozen
To a squadron commander	6 dozen
To a captain of a 1st rate ship	5 dozen
To a captain of a 2nd rate ship	4 dozen
To a captain of a 3rd rate ship	3 dozen
To a captain of a 4th rate ship	2 dozen
To a captain of a 5th rate ship	1.5 dozen
To a captain of a 6th rate ship	1 dozen

Source: ADM 30/44, p. 21.

The men's rations were sufficient in terms of quantity, and it seems that they were of good quality too, judging by the very small percentage that spoiled. A report written on 15 February 1758 by the Victualling Board Commissioners for the Lords of the Admiralty, concerning the thirteen most commonly supplied products, clearly indicated in what quantities these supplies had been delivered, and how much wastage had occurred between 1 January 1749 and 31 December 1757, and consequently made it possible to calculate accurately the percentage of loss (see Table 3.6).

23 Ibid.

24 Victualling Board to the Admiralty, dated December 1703, in Merriman, *Queen Anne's Navy*, extract no. 10, 268–9.

25 There are no accounts records to be found which would prove that the men did benefit from the system. Fresh produce was bought directly by the crew or by the ship's purser and neither was under an obligation to furnish the Navy administration with a receipt.

Table 3.6 Victuals which were discarded between 1 January 1750 and 31 December 1757

Produce	Quantities supplied	Perished quantities	Percentages
Bread	54,642,437 lbs	185,761 lbs	0.34
Beer	110,049 tuns	1,062 tuns	0.97
Brandy	351,692 gals	None	0
Beef	4,498,486 lbs	2,653 lbs	0.06
Pork	6,734,261 lbs	2,217 lbs	0.03
Peas	203,385 bushels	1,201 bushels	0.59
Flour	6,264,879 lbs	20,858 lbs	0.33
Suet	809,419 lbs	895 lbs	0.11
Raisins	705,784 lbs	702 lbs	0.10
Porridge (oatmeal)	138,504 bushels	1,214 bushels	0.88
Vinegar	290,863 gals	None	0
Olive oil	71,668 gals	322 gals	0.04
Stockfish	166,943 lbs	13,043 lbs	7.81

Source: British Library Stowe MS 152, f. 130. I am grateful to N.A.M. Rodger (*Command of the Ocean*, 305 and 697, n. 1), J. Macdonald ('A new myth of naval history?', *International Journal of Maritime History*, XXI, 171) and Michael Duffy for pointing out transcription errors in Gradish, *Manning of the British Navy*, 144, previously used by the author. The corrections strengthen the case.

With the exception of fish, the items each show a loss of less than 1 per cent. These statistics are undoubtedly trustworthy for it was not in the officers' interests to minimise losses. If anything, the pursers tended to exaggerate losses because they were then in a position to make a profit since they were responsible for placing the orders to replace the foodstuffs.

These statistics therefore reflect the high standards of food preservation that the Victualling Board had achieved. The percentages for meat loss (0.06 per cent for beef, 0.03 per cent for pork) clearly illustrate the progress which had been made since the War of the Spanish Succession, in which salt meat had been particularly badly preserved: 'the salting of beef and pork was done crudely; a high proportion had to be condemned after being on board but a few months'.[26]

As soon as that war had ended the Commissioners started to look into different food preservation techniques, and the two methods chosen were imposed by the administration on 27 January 1716 (see Appendix 6). One was for the salting of beef, the other for that of pork.[27] These methods proved

26 Merriman, *Queen Anne's Navy*, 250.

27 Instructions given to Samuel Rowe, clerk of the cutting house at the port of London (27 June 1716), in Baugh, *British Naval Administration*, extract no. 5, 413–14.

to be highly satisfactory and were re-approved in 1736, 1764, 1766 and 1784. But satisfaction with a particular method did not imply complacency, and the Commissioners continued to be watchful for any innovation that improved the meat curing process. A number of ideas were tested, especially during the 1739–48 war,[28] but apparently without encouraging results.[29] Only one new method, using brine and suggested by Thomas Lowndes,[30] was to have important consequences. Lowndes's saline solution was considered to be as good if not better than that which had previously been obtained by using French bay salt, and this new brine would be used systematically from 1754 onwards for the salting of meat.[31]

In fact, all produce was treated with great care. First of all, only the best-quality products were purchased. Secondly, great care was taken in processing, with the promulgation of certain rules and regulations, some of which went so far as to state the dimensions of each container.[32] In this way, and contrary to popular belief, the Commissioners of the Victualling Board did succeed in obtaining impressive results in terms of food preservation in the mid-eighteenth century. Nothing escaped their scrutiny, including the conservation of water. In 1756 the Commissioners experimented with a procedure invented by Dr Hales that was supposed to stop the stagnation of water. The process involved blowing air into the water using a series of small bellows. Mr Yeoman, the Victualling Board's inspector, judged the machine to be efficient and a great many orders were placed for it during the Seven Years War.[33]

Innovations or Food Revolution?

Because lemons were not systematically introduced into the diet before the very end of the eighteenth century, it is too easily forgotten that the Victualling Board made a huge effort to protect seamen against scurvy. Indeed, the Seven Years War should be considered as a period in which huge and inno-

28 Baugh, *British Naval Administration*, 427–8.

29 Victualling Board to the Admiralty on the subject of the method suggested by William Guy (3 July 1745), in Baugh, *British Naval Administration*, extract no. 44, 445–7.

30 Lowndes was a Cheshire landowner, clerk to the Treasury and colonial sinecurist turned writer on trade matters, See J.A. Cramb and T.O. Bickham, 'Thomas Lowndes, bap. 1692–1748', in *Oxford Dictionary of National Biography* (Oxford, Oxford University Press, 2004).

31 The process did not obtain as good results in the salting of fish. For more on this method, see PRO, ADM 110/19, 367–8.

32 PRO, ADM 30/44, 20.

33 See for example PRO, ADM 110/18, 27 April 1757; ADM 110/19, 28 February 1759; ADM 110/20, 16 March 1761.

vative changes were taking place in terms of diet as a response to the fight against scurvy. A number of physicians of that time – for example, James Lind, who had published his well-known *Treatise of the Scurvy* in 1753 – drew the Board's attention to certain types of food considered to prevent the illness. As a consequence, these foodstuffs were gradually introduced and eventually included in the composition of all menus – a fact which has not been sufficiently recognised, especially in light of the excellent results this had in the fight against scurvy.

How much was known or understood about scurvy in the eighteenth century? The answer to this question provides invaluable insight into the contribution that the Seven Years War brought to the fight against this diet-deficiency disease. Speculation at the time as to what caused the disease was the subject of fierce debate between those who believed that the illness was fundamentally linked to problems of diet, and others – Richard Mead[34] and William Cockburn,[35] for example – who, while recognising the beneficial aspects of certain fresh fruits and vegetables, were nonetheless of the opinion that scurvy was the direct result of the harsh living conditions that the seamen had to endure, and in particular the stale air within the enclosed decks of the ships. By the mid-eighteenth century it had become common practice to saw open windows in the hull in order to allow the air to circulate more freely. This was illustrated, for example, by entries in Admiral Anson's log book during his circumnavigation of the world between 1740 and 1744.[36] The comparison of various voyages undertaken during this epic period in maritime history – a period which was particularly badly hit in terms of sickness – does allow particular attention to be paid to the theory, which was beginning to emerge from the mid-eighteenth century, of diet as an explanation for scurvy. It was noticed that once the men had reached dry land and could consume fresh produce (*refreshments*, to use the vocabulary of the period), they would recover fairly quickly. The conclusion, therefore, was that the illness was caused by bad digestion due to over-consumption of salted food.[37]

To this end, elixir of vitriol had already been added to medicine chests as of the 1739–48 war, because it was believed to dissolve salt. The addition of mustard[38] and cider to the rations of seamen who were setting off on long

34 R. Mead, *A Discourse on the Scurvy* (1749), and published in *The Medical Works of Richard Mead M.D.* (Edinburgh, 1763).

35 W. Cockburn, *Sea Diseases* (London, 1696).

36 Ch. Buchet, *Le Tour du Monde de l'Amiral Anson* (Paris, France-Empire, 1992).

37 This still continued to be the point of view taken by Henri-Louis Duhamel de Monceau at the beginning of the nineteenth century.

38 Cf. PRO, ADM 111/49. Detailed list of victuals loaded onto the ship *Southsea Castle* in readiness for departure to the West Indies, 26 December 1759.

voyages,[39] plus the half-pint of vinegar which was added by the order of 3 July 1740, followed the same lines of thought.[40]

However, one of the three most important changes to take place in victualling during the Seven Years War was the decision to substitute fresh meat for salted beef and pork. Even though this decision was, as it happens, a step in the wrong direction since meat does not provide a significant vitamin C intake, it does deserve to be mentioned if only to underline the Victualling Board's efforts to try to eradicate scurvy – efforts which have been unjustly ignored by maritime historiography.

On 17 January 1747, impelled by Admiral George Anson – a Lord of the Admiralty who had also recently been appointed commander of the western squadron – it was decided that victualling with fresh meat (which had been limited to two distributions when the men were stationed in the home ports since the orders of 12 December 1704 – cf. the *petty warrant* ration)[41] should be doubled for crews who had been afflicted by scurvy or any other disease, on their return: 'The ships companys of cruising ships should be supply'd with fresh meat and greens four days in a week when they come into port afflicted with the scurvy or other illnesses.'[42] On 10 February 1748, this measure was extended to all ships, and extended geographically to include all Irish ports.[43]

Regular victualling with fresh meat was developed still further during the Seven Years War and a truly global system of organisation was put into place. Contractors who were responsible for victualling the Royal Navy in outlying zones were themselves put under the constraint of finding fresh meat. In other words, they had to ensure that they could fulfil any orders for fresh beef that were presented to them, and at the price that had previously been agreed upon. There were only two locations in which the local contractors were exempted from this rule: Madeira, and Barbados and the Leeward Islands,[44] because these areas suffered from an almost total dearth of cattle, resulting in too high a financial outlay.[45]

39 PRO, ADM 111/33, 17 July 1747.

40 PRO, ADM 110/12, 20 February 1741.

41 Merriman, *Queen Anne's Navy*, 251.

42 An Admiralty directive, cited in Baugh, *British Naval Administration*, 380. Anson, whose squadron had been devastated by scurvy during his famous voyage around the world in 1740–4, was made a peer after the western squadron's victory off Cape Finisterre in 1747.

43 PRO, ADM 3/58, 10 February 1748.

44 Contracts concerning Madeira of 2 July 1756, 1 November 1758 and 15 March 1762 in PRO, ADM 111. Contracts concerning Barbados and the Leeward Islands of 11 August 1756, 28 April 1758 and 7 March 1760 in PRO, ADM 111.

45 Victualling Board to the Admiralty, 21 March 1735, in Baugh, *British Naval Administration*, extract no. 29, 433–4.

British ships putting in at Lisbon, Oporto, Gorée, Jamaica and New York and then eventually at Louisburg, Quebec and Havana could thus take on board rations of fresh meat. Due to the very large number of ships docking in Jamaica and the insufficient quantity of livestock, this particular region obtained a special contract concerning the victualling of fresh meat. In fact the supplier benefited from a highly advantageous clause in that the agreement was drawn up for a minimum period of four years, and a full year's notice was required in order to terminate the contract.[46]

The Victualling Board Commissioners did not limit their efforts to the distribution of fresh meat rations in the ports, however. In order to fully exploit the beneficial effect that meat was supposed to have in the fight against scurvy, and in order to be able to do so on long-haul voyages that did not have an intermediary port of call, it was decided to load live oxen that could be slaughtered as and when the need arose. The quantity of cattle[47] and sheep on board a vessel sometimes reached huge numbers, as N.A.M. Rodger points out. As it got underway for the East Indies, the *Elisabeth*, a third rate ship, was carrying no less than seventy-one head of cattle; and the vessels under the command of Sir Edward Hawke blockading Brest each took on board one dozen head of cattle and about forty sheep.[48] It goes without saying that if the same efforts had been made in the victualling of leafy green vegetables rich in vitamin C, scurvy would have been wiped out far sooner.

The second big change to take place in this period was the introduction of fresh vegetables. In December 1747, John Troughton suggested to the Admiralty that fresh vegetables should be systematically distributed with the meat rations. For a while, the Victualling Board's Commissioners contented themselves with leaving the decision to carry vegetables to the discretion of the captains and pursers of each ship.[49] But Admiralty directives issued on 10 February 1748 now urged captains to ensure that the pursers provided what they considered to be sufficient quantities of greens and roots. Nonetheless, the quantity of these supplies remained discretionary.[50]

Whilst these measures did not systematically introduce fresh vegetables into daily rations, in the instances where they were introduced they did immediately reveal encouraging results. As a consequence, there was a growing demand for the regular consumption of fresh vegetables throughout the Seven Years War. The measures had the firm support of a number of influen-

46 Contracts concerning the victualling of fresh meat in Jamaica of 8 June 1756, 18 February 1757 and 4 January 1761 in PRO, ADM 111.

47 Cattle slaughtered for consumption were always oxen – castrated male bovines.

48 Rodger, 'The victualling of the British Navy in the Seven Years' War', 44.

49 Baugh, *British Naval Administration*, 380–2.

50 PRO, ADM 3/58, 10 February 1748.

tial and highly determined Admirals. For example, Admirals Anson, Hawke and Boscawen insisted that the western squadron, which was operating at the entrance to the Channel and in the Bay of Biscay, should be regularly victualled by supply ships, not only with meat but also with fresh vegetables.

The third innovation was the introduction of portable soup. This was a soup based on meat stock, usually mutton, which had been solidified into cubes. A Mrs Dubois of London perfected the process.[51] (The National Maritime Museum is still in possession of one of these cubes, dating from some time between 1765 and 1779, and according to an analysis carried out on it in 1928, it was still edible.[52]) The soup and its method of conservation[53] soon came to the attention of the newly appointed medical staff[54] of the Sick and Wounded Board. It was realised that the soup would undoubtedly help sick seamen who could not take in solids to keep up their strength, and indeed to better nourish themselves since the meat in the soup was hardly salted.[55] Even though originally destined for the sick, the ensuing enthusiastic reports written by the ships' surgeons[56] meant that soon after the end of the war portable soup became a regular feature in every seaman's rations. The Victualling Board quickly established a factory in Deptford that produced this foodstuff. The ADM. 111 series lists all the contracts that the Commissioners now began to draw up on a quarterly basis with meat suppliers, from 24 January 1757.[57] The dealer who won the contract had to agree to supply whatever quantity of mutton might be demanded of him. The outstanding benefit of portable soup, however, was the fact that it contributed to the widespread use of vegetables, as it soon became customary to include vegetables in the preparation of the soup. This preparation had become standard practice by the end of the war and was clearly beneficial in the fight against scurvy.[58]

Victualling and Strategy: The Example of the 'Western Squadron'

In *The Manning of the British Navy during the Seven Years War*, Stephen F. Gradish quite rightly claims that 'the Admiralty's plan to provide fresh vict-

[51] Gradish, *The Manning of the British Navy*, 159.

[52] Lloyd, 'Victualling of the fleet in the 18th and 19th centuries', 12.

[53] PRO, ADM 98/5, 24 August and 11 September 1756.

[54] Gradish, *The Manning of the British Navy*, 20–1.

[55] J. Lind, *An Essay on the most effectual means of preserving the Health of Seamen* (London, new ed. 1774).

[56] PRO, ADM 98/7, 4 January 1758 and 21 January 1759; ADM 98/8, 31 December 1760.

[57] PRO, ADM 111/45, 24 January 1757.

[58] See, for example, PRO, ADM 110/21, pp. 422–4, 16 January 1764.

uals at sea during the latter half of 1759 opened a new chapter in the history of naval victualling'.[59] This improvement in logistics was undoubtedly a determining factor in British success.

The initial stimulus came from a shift in British strategy in the spring of 1759. Aware of the Navy's superiority in terms of sheer numbers, the First Lord of the Admiralty, Lord Anson, set about trying to maintain a close observation of the French Navy in its ports, thus inaugurating the tactic known as *close blockade*. So, rather than cruising out in the Bay of Biscay, the western squadron – which since April 1747 had been under orders to interrupt all French communications – now received orders to station itself just a few miles off the French coast near Ushant in order to prevent all French coastal navigation and thus thwart the concentration of French forces seeking to make a landing on British soil.[60]

In fact, it was the western squadron's distance from its victualling bases in the summer of 1759, while it was stationed off Brest and Quiberon, which prompted the Victualling Board to set up a connecting logistical system. Basically, allowing the ships to return to England for victualling purposes, even in rotation, would have weakened the concentration of vessels off the French coast. The Victualling Board therefore had to establish a system of links. The operation was entrusted to Mr Ommanney, the Victualling Board's Agent at Plymouth, who ensured that the western squadron received from that port all the foodstuffs that were required in the ration list. Furthermore, in order to ensure sufficient revictualling, and especially to facilitate the supply of fresh meat, quarterly contracts were established with suppliers which stipulated that the ships could replenish their supplies whenever necessary at Torbay – which was the nearest British fleet anchorage to the theatre of operations.[61]

These measures clearly did not eradicate the appearance of scurvy after six weeks at sea, which left Admiral Edward Hawke with the fear that he would have no option but to end the blockade. On 23 July, in a message addressed to the Admiralty, the commander of the western squadron therefore insisted on the immediate delivery of live animals.[62] Their Lordships, under the leadership of Anson, were ready to do whatever it took to rectify the situation. In less than a fortnight, six ships carrying live animals set sail in the direction of Sir Edward Hawke's squadron. Possibly of even greater importance as a reflection of the evolution of ideas in food management was the inclusion in this cargo, on the initiative of the Commissioners of the Victualling Board,

59 Gradish, *The Manning of the British Navy*, 168.

60 J.S. Corbett, *England in the Seven Years' War: A Study in Combined Strategy* (London, 1907).

61 Cf. the ADM 111 series in the archives.

62 PRO, ADM 1/92, Hawke to Admiralty, dated 23 July 1759.

of fresh vegetables, especially cabbages, carrots and turnips.[63] Within a few weeks, this one-off situation had been transformed into regular victualling of the squadron at sea with fresh produce. A sloop called the *Catwater* was responsible for victualling the squadron with vegetables and potatoes,[64] and there was a weekly delivery of livestock and sheep.[65]

The vitamin intake from a regular diet of fresh vegetables soon began to have beneficial effects. Scurvy may not have been totally eradicated, but at least it was now controlled to such a degree that it no longer had a detrimental effect on naval strategy. Indeed, when the Battle of Quiberon Bay took place on 20 November 1759, James Lind noted that the British seamen: 'enjoyed a most perfect and unparalleled state of health'.[66] He continues:

> The squadron consisted at most times of above twenty ships of the line, and ten or more frigates, in which were embarked about fourteen thousand men. On the day of action, many of these ships and men had been above six months from Spithead; notwithstanding which, there was not among them twenty sick in all.... It was hardly ever known before, that ships could cruise in the Bay of Biscay much above three or four months at a time; without having their men afflicted by the scurvy. An exemption from which was entirely owing to this fleet having been well supplied with fresh meat and greens.[67]

The following year, when the squadron went to sea again under the command of Admiral Boscawen, regular victualling in fresh vegetables and live animals had become normal procedure. The contract for supplies was granted to an independent contractor, Mr John Biggin, who was one of the biggest traders of the time. For financial reasons, Mr Biggin established his operations from Ireland, and three ships, the *Rose*, *Lyon* and *Goldsburry*, regularly carried supplies of live animals, fodder with which to feed them, potatoes and onions[68] from Cork and sometimes from Kinsale to the squadron.[69] In order to be prepared for all eventualities, Mr Ommanney at Plymouth, who continued to supply the squadron with the foodstuffs that made up rations, was also instructed to include cabbages, onions and other vegetables in each load.[70]

63 PRO, ADM 111/49, 8 August 1759.

64 PRO, ADM 2/83, 20 August 1759.

65 PRO, ADM 111/49, 4 August 1759.

66 Lind, *An Essay*, 215.

67 Lind, *An Essay*, 215.

68 Cabbages are also listed from time to time: for example in PRO, ADM 111/52, 27 July 1761.

69 PRO, ADM 111/50, 18 July and 5 September 1760.

70 PRO, ADM 111/50, 18 July 1760. See also ADM 111/52, 27 July and 4 September 1761.

It proved more difficult to keep scurvy at bay during the winter months when the bad weather conditions rendered it more difficult to regularly supply the ships with fresh foodstuffs. Nonetheless, the British successfully maintained the blockade tactic until the end of the war.[71] In 1761, John Biggin deployed no fewer than sixteen ships per trip in order to fulfil his contractual duties.[72] The Board's Commissioners constantly scrutinised Biggin's services and did not hesitate in summoning him if they felt the slightest doubt about the quality of the victuals he provided,[73] or requiring him to increase the quantity of fodder he sent so that the live animals already on board the squadron's ships could benefit from it too.[74]

The role of vegetables in the prevention of scurvy became so important that Admiral Boscawen, who commanded the western squadron in 1760, ordered that they be grown on a small island that he had seized just off the coast at Vannes.[75] Commodore Keppel followed suit the following year when he, in turn, captured the island of Belle-Ile.[76]

The changes that have been cited in this chapter regarding the composition of rations, and especially regarding the generalised inclusion of fresh meat and vegetables, had extremely beneficial results in terms of the health of the seamen. In the last years of the war, scurvy would no longer be a disrupting factor in the operations and movements of the squadrons, and N.A.M. Rodger has also shown that there was a sharp drop in death and morbidity rates between 1757 and 1762.[77]

This improvement also took place thanks to First Lord of the Admiralty, Lord Anson, and his unceasing preoccupation with the crucial issue of 'the good condition of the men'. Through his impetus, the Sick and Wounded Board, which was responsible for establishing the contents of food rations and which, since its conception in the seventeenth century, had been involved with questions of health, now became a professional body, with the nomination of two physicians to the posts of two of its four Commissioners, namely, James Maxwell and Henry Tom.[78] The Sick and Wounded Board continued to develop its hospitals – Haslar Hospital, near Gosport, would be expanded

71 The Victualling Board's Commissioners did not terminate Mr Biggin's victualling contract until 29 November 1762. PRO, ADM 111/54, 29 November 1762.

72 PRO, ADM 111/52, 4 September 1761 and ADM 111/53, 6 January 1762.

73 PRO, ADM 111/52, 16 October 1761.

74 PRO, ADM 111/54, 10 September 1762.

75 Aspinall-Oglander, *Admiral's Wife*, 283, cited in Gradish, *The Manning of the British Navy*, 169.

76 PRO, ADM 1/91, Keppel at the Admiralty, 6 November 1761.

77 Rodger, *The Wooden World*, 98–105.

78 Gradish, *The Manning of the British Navy*, 21.

until 1761, and Stonehouse Hospital, near Plymouth, was constructed between 1758 and 1761 – and it also kept a check on the competence of the medical staff and organised inspections.[79]

In conclusion, the parallel and complementary actions of a medical and administrative nature undertaken by the Commissioners of the Sick and Wounded Board and the Victualling Board put the Royal Navy, from the Seven Years War onwards, in a position to contain the problem that had most often determined the success or failure of its operations: the health factor.[80] The next section of the book will examine the administrative infrastructure established by the Victualling Board which enabled this.

79 Rodger, *The Wooden World*, 109–12.

80 Ch. Buchet, *La Lutte pour l'Espace Caraïbe* (Paris, Librairie de l'Inde Editeur 1991), II, 693–818.

PART TWO
THE BASES

4

The Victualling Board's Principal Bases

The Victualling Board's principal bases were under the direct jurisdiction of this Admiralty office responsible for organising the different operations relating to victualling. An Agent of the Victualling Board was in charge of personnel, the number of which varied according to the situation of the base, and he managed victualling activities in close liaison with the Board Commissioners based in London.

The number of bases directly under the jurisdiction of the Victualling Board noticeably decreased during the first three conflicts in the eighteenth century (see Table 4.1). There were eleven bases during the War of the Spanish Succession, but then, with the abandonment of victualling at Lisbon and Harwich in 1713,[1] there were only nine bases left to be directly victualled by the Victualling Board during the 1739–48 war. During the Seven Years War, this number fell to six. The fortress at Port Mahon in Minorca fell at the outbreak of hostilities, which led to the court martial, death penalty and subsequent execution of the British naval commander, Admiral Byng, who had failed to relieve the fortress. Victualling for Jamaica and Kinsale[2] was entrusted to a private contractor who was responsible for meeting all the needs of any of the ships, according to an established procedure. This will be examined more closely in Chapter 6, which is devoted to victualling in the secondary ports and outlying centres.

This evolution clearly shows that, as far as possible, the Victualling Board intended to rely on the competitive commercial system for victualling the Navy's far-flung ships. In fact, with the exception of Gibraltar, the Victualling Board preserved the victualling only of the primary bases which, owing to the very nature and importance of their activity, it was felt should not be placed under the control of a private contractor.

The five primary home bases directly dependent on the Victualling Board were part of a geo-strategic plan. The port of London was the nerve centre of this plan, through which passed the essential foodstuffs required by the

1 J.G. Coad, *The Royal Dockyards 1690–1815* (Aldershot, Scolar Press, 1989), 10.

2 Victualling at Cork eventually replaced that carried out at Kinsale, where the bar at the mouth of the River Brandon prevented access by the increasingly large ships of the line.

Table 4.1 Bases under the direct responsibility of the Victualling Board

War of the Spanish Succession: 11	
London	Kinsale
Portsmouth	Lisbon
Plymouth	Gibraltar
Dover	Port Mahon
Chatham	Jamaica
Harwich	
War of 1739–48: 9	
London	Kinsale
Portsmouth	Gibraltar
Plymouth	Port Mahon
Chatham	Jamaica
Dover	
Seven Years War: 6	
London	Chatham
Portsmouth	Dover
Plymouth	Gibraltar

other bases under the Board's auspices. Included in the port of London were the warehouses at Deptford, Woolwich and Redhouse, which were situated downstream of the City of London on the Thames. During the Seven Years War, the yards at Chatham and Sheerness were no more than dependencies of the port of London. Annual victualling personnel statistics issued throughout the war only mention four geographical centres within the country: London, Portsmouth, Plymouth and Dover.

The loss of Harwich as a victualling port, and the linking of Chatham and Sheerness to London, illustrate the degree to which the east coast was considered unimportant in the overall geo-strategic plan. The time of the Anglo-Dutch wars was over, a time in which it had been necessary both to stop enemy raids on the Thames and to take advantage of well-sheltered ports in the face of attacks from the United Provinces.[3] It was then that Chatham had its heyday, since its position on the River Medway rendered it well protected and admirably situated; Sheerness and Harwich had likewise added to this defence system as they faced east, towards the North Sea. Dover, Portsmouth and to an even greater extent Plymouth were little more than secondary ports in the seventeenth century.[4] The development of commercial links with

3 D.C. Coleman, 'Naval Dockyards under the Later Stuarts,' *Economic History Review*, 2nd series, 6 (1953), 134–55.

4 For information on the harbour infrastructure of Plymouth at the beginning of the eighteenth century, see Michael Duffy's excellent study, 'The creation of Plymouth

America and an increase in French maritime strength would disrupt this situation from the end of the seventeenth century. The north-south line of naval ports on the east coast facing Holland was gradually replaced, during the eighteenth century, by an east-west line of bases along the Channel. The south coast of England, facing France and closest to the transatlantic routes, progressively became the main zone of maritime activity. The silting up of Chatham[5] and its difficulty of access due to its position on the river, which was less of a problem in the seventeenth century when the ships were of lesser tonnage, no doubt contributed to this change.

London, during the eighteenth century, was more than ever at the heart of victualling, due to the importance of its agricultural market. However, the arc of the circle of primary bases which the capital supplied was displaced to the south: to Portsmouth and Plymouth, where the squadrons were armed, maintained and resupplied, which continued to rise in importance, while Dover, which was extremely well placed as a port for ships operating in the narrows of the Channel, maintained its level of activity.[6]

The development of the infrastructure of these victualling yards which depended directly on the administration of the Victualling Board will be shown in the first section of this chapter, before examining the importance of each port in the victualling of the Royal Navy, based on the yard pay lists in ADM 113 in the Public Record Office which contain details of the wages of the personnel in these bases.

The Development of an Infrastructure of Port Facilities

1701–38: A long period of inactivity

The Victualling Board's lack of adequate installations in the ports directly under its jurisdiction was sorely felt during the War of the Spanish Succession. For example, in correspondence addressed to the Lords of the Admiralty on 31 March and 20 June 1703, 8 August, 21 October and 30 November 1704, 20 October 1705, 20 September 1706 and 12 March 1707,[7] the Commissioners insisted on the paramount need to build, in the ports themselves, breweries, bakeries and sufficient stores in which to stock the goods, in order to control the abuse and fraud which the contractors were prone to. A letter

Dockyard and its impact on naval strategy', in *Guerres Maritimes (1688–1713)*, Vincennes, Service Historique de la Marine, 1996, 245–74.

5 Coad, *The Royal Dockyards*, 13.

6 Baugh, *British Naval Administration*, 436.

7 Victualling Board to the Admiralty (21 March 1711), in Merriman, *Queen Anne's Navy*, 298–300.

addressed to the Lords of the Admiralty on 21 March 1711 gives the precise reasons for this:

> The said buildings have afforded very great advantages to the service, as well as in the goodness of the bread and beer (for which we should have been able to set the contractors a standard) as in preventing their taking advantages by combinations, and imposing their own prices on every little rise of the market. The having sufficient store rooms would also prevent the sending of bread and beer directly from the contractors on board the ships, which there is now in great measure a necessity of doing; whereby the quantity and quality cannot be so constantly viewed by our officers as it ought to be.[8]

Some years previously, in March 1707, the Lord High Admiral himself, in a letter addressed to the Queen's Council, had already, and without success, requested that the Crown have a bakehouse and brewhouse built in Plymouth. The letter stated that the beer procured in this port from private suppliers was priced 25 per cent above that demanded at other yards. Furthermore, the beer was so bad that it had been shown on numerous occasions to damage the men's health, and the Lord High Admiral had to accept the complaints presented to him by Admiral Cloudesley Shovell, regarding the very poor quality of the beer and bread supplied from this port. The cost of such buildings was at the time estimated at £9,754.[9]

Following the end of the Spanish War of Succession and during the long period of peace that followed, up to 1739, if one excludes a few minor operations in 1727, the Victualling Board Commissioners consistently requested that the necessary infrastructure be constructed in the principal naval ports. However, a combination of debt and the paltry credits granted to the Navy[10] did not allow such badly needed projects to get underway.

In February 1717[11] the Commissioners warned that lack of financial resources was such that it did not even allow for the upkeep of existing buildings. In July 1718 the Commissioners undertook a tour of inspection of the different ports and observed that the buildings at Chatham were so old and dilapidated that they were not worthy of repair: 'such old buildings as would be ready to fall to pieces almost as soon as the work was done'. It seemed

8 Ibid.

9 Report of the Lord High Admiral to the Queen's Privy Council (18 March 1707), in Merriman, *Queen Anne's Navy*, 280–1.

10 For example, the Commissioners of the Victualling Board wrote to the Admiralty on 12 November 1718, saying that 'The said allowance even in the best and cheapest years was not found sufficient to answer the real charge thereof' (document reproduced in Baugh, *British Naval Administration*, 423–4).

11 Victualling Board to the Admiralty (19 February 1717), in Baugh, *British Naval Administration*, 417–18.

to them 'highly necessary' that a new set of buildings be erected despite the financial difficulties of the Crown:

> We are sensible that the want of proper conveniences there have been already no small addition to the debt, and it is no less certain that a fruitless endeavour to repair the old buildings will be a further addition, especially considering the detriment that the provisions must receive in such buildings, where they cannot be duly preserved either from the weather or embezzlements.[12]

In 1720, the Commissioners announced what works needed to be carried out in the ports of Plymouth and Portsmouth.

It was war against France in 1689, fought in the context of the War of the League of Augsburg, which led to the transformation of Plymouth, from the simple, small commercial port it had been since the thirteenth century, into a military base. In 1690 work began which culminated in the establishment of the first buildings on the east bank of the Hamoaze, westward of the town where the victualling warehouses still remained adjacent to the town citadel. During the War of the Spanish Succession, the extensive development of naval activities operating from the dockyard on the Hamoaze made it clear that the infrastructure under the Victualling Board's authority was inadequate. Letters written to the Admiralty in 1720 after the conflict refer to the Commissioners' proposal to purchase an estate near Empacombe, across the Hamoaze from the new dockyard, which belonged to a private individual, Mr Edgcumbe. It had the advantage of being more favourably situated in relation to the predominant winds.[13] The project was put on file, and it was only in 1729 that the building of the brewery that had been requested for so many years was finally given consideration.[14] Work on the brewery began in 1733,[15] but its production capacity of 80 tons per week proved insufficient to meet wartime needs.[16]

Inspections carried out in Portsmouth also revealed the deplorable state of buildings belonging to the Victualling Board in this yard: in their letter of

12 Victualling Board to the Admiralty (30 July 1718), in Baugh, *British Naval Administration*, 422–3.

13 Victualling Board to the Admiralty (15 January 1720), in Baugh, *British Naval Administration*, 425–7. See also the report of the Lord High Admiral to Her Majesty's Privy Council (18 March 1707), in Merriman, *Queen Anne's Navy*, 280–1.

14 Coad, *The Royal Dockyards*, 282.

15 Victualling Board to the Admiralty (10 January 1733), in Baugh, *British Naval Administration*, 432–3.

16 P. Mathias, *The Brewing Industry in England 1700–1830* (Cambridge, Cambridge University Press, 1959), 201–3.

18 May 1720, the Commissioners used the term 'ruinous condition'.[17] They particularly drew the Admiralty's attention to the need to build a brewery in Portsmouth,[18] but this was not to see the light of day until the middle of the century.[19]

Problems posed by the port of London were no less serious. The London victualling headquarters was situated at Tower Hill, on the site formerly occupied by the Abbey of St Mary Graces. After the demolition of the abbey during the Reformation, facilities were built by the Navy Board, which at the time directed the Victualling Board. They were made up of offices, warehouses and ovens for baking flour into biscuits. The relocation of the Navy Board to Seething Lane in 1654 left the Victualling Board in charge of the premises, and soon a cooperage was built. Partially destroyed by fire, the main buildings had to be rebuilt in 1701. Another group of buildings housed the cuttinghouses and spice storerooms, while a brewery was built nearby at St Catherine (today known as St Catherine's Dock). In 1704 further stores had to be added for biscuits, beer and meat.[20] To complete these facilities in London, since 1698 the Victualling Board had been leasing mills at Rotherhithe, which was only half a mile southeast of London on the banks of the Thames. This was costing the Board £135 per annum.[21]

The location of the London victualling station at the heart of the capital limited its possibilities for expansion. Its Thames-side wharf soon proved inadequate to efficiently carry out the loading of an increasing quantity of goods. Hemmed in by Ordnance Board buildings independent of the Admiralty, the portion of the wharf the Victualling Board had at its disposal was far too narrow; so only two cranes could be positioned there. From 1724 onwards, all requests placed by the Commissioners of the Victualling Board for the attention of the management at the Ordnance Board, calling for an extension of wharf space, were refused because the Ordnance Board also had a pressing need to use the wharf for its own loading purposes.[22]

17 Victualling Board to the Admiralty (18 May 1720), in Baugh, *British Naval Administration*, 427–8.

18 PRO, ADM 110/8, 27 June 1721.

19 Mathias, *The Brewing Industry*, 201–2.

20 Merriman, *Queen Anne's Navy*, 248–9.

21 The cost of this rental was reduced to £100 in August 1713 by a lease fixed at twenty-one years (Admiralty to the Navy Board, dated 2 August 1713, in Merriman, *Queen Anne's Navy*, 304).

22 The history of this contentious issue is traced in the letter from the Commissioners of the Victualling Board dated 9 December 1740 in ADM 100/12.

1739–48: The sudden awakening

The 1739–48 war was a period of major importance in the history of the British naval bases. Victualling deficiencies at the outbreak of hostilities, due, as has already been seen, to an inadequate infrastructure, were such that the Crown could no longer postpone the construction work that the Victualling Board had been demanding for so many years. Consequently, within a few years, and following a period of great inactivity, the Royal Navy victualling yards would be completely transformed.

In London, the Victualling Board dealt with its most urgent issues by leasing warehouses and ordering six ovens to be built to meet the great demand for bread and biscuits.[23] It also had to consider the construction of another victualling complex, since the Tower Hill site could not be expanded. As a result, on 14 May 1742, the Commissioners made a request to the Lords of the Admiralty that the Crown purchase the property known as Redhouse, belonging to Sir John Evelyn. On these eleven acres, situated between Deptford and Cuckold Point, there already existed a few warehouses and, above all, there was an 800-foot wharf where three or four ships could be loaded simultaneously. The Victualling Board Commissioners did not hide the fact that it would be necessary to carry out large-scale repairs and construct numerous buildings in order to make the yard operational:

> It would require very large sums of money to render the whole completely useful, but how far that expense is to weigh against the great advantages the service would reap by it when so completed, or whether it would not be more advisable to be at any extraordinary charge of that kind than to hazard the frustrating any one great expedition or design by delay or otherwise which might unfortunately happen from the want of conveniences.[24]

The project was submitted to the Royal Council on 16 August 1742. The latter rejected the purchase of the land, but since Evelyn then agreed to lease it to the Crown for £500 per annum, the Victualling Board was able to take possession at the end of spring 1743.[25]

The construction work began forthwith. First of all, the Commissioners ordered a large warehouse to be built, intended to contain all the produce which had been stored in the different warehouses rented for this purpose around Tower Hill. A mill was also built to grind oats.[26] In 1747 a wall was

23 Victualling Board to Admiralty (12 September 1739), in Baugh, *British Naval Administration*, 435.

24 Baugh, *British Naval Administration*, 435.

25 For more information on the development of Redhouse, see Baugh, *British Naval Administration*, 433–5.

26 PRO ADM 110/13, 8 September 1743.

built to encircle the Redhouse property, plus a small-scale cooperage, built to refurbish used casks. In the same year it was decided to build another bakery big enough to meet the needs of the naval yards of the port of London, as a fire had destroyed the bakery at Rotherhithe. The cost of this work was estimated at £4,000.[27]

If Chatham was only enlarged by one small warehouse during the hostilities,[28] this reflects the lesser role that the port was to play due to the change in the strategic axes of operation to Portsmouth and especially Plymouth. Facilities at these two naval ports were greatly expanded due to their favourable position on the Channel coast and their proximity to Atlantic routes. Portsmouth did not yet have a bakery belonging to the Crown, but at the outbreak of hostilities the victualling yard was completed by building six ovens, as had previously happened in London, in order to meet the heavy demand for bread and biscuits.

After London, Portsmouth became the busiest victualling base in Britain during this war,[29] and in 1744 a mill was built since the Commissioners preferred to process grain directly rather than rely on contractors to deliver flour.[30] In April 1741 the construction of a bakery also began in Plymouth.[31] Changes needed in this port regarding the buildings belonging to the Victualling Board were carried out at the beginning of 1745, in light of the increase in Plymouth's traffic and the importance of its duties. In view of the pitiful state of its main buildings, a new wharf was constructed on which were built various storehouses and shelters, and, in addition to the bakery constructed at the beginning of the hostilities,[32] it was decided to build another, because 'bread ... bought from private contractors is not of such good quality as that made in his Majesty's bakeries.'[33]

27 PRO ADM 110/15, 9 March, 6 April, 6 May, 22 June 1747; ADM 110/33, 22 April 1747. These references, as well as the one in the preceding footnote, have been taken from Baugh, *British Naval Administration*, 435.

28 PRO ADM 110/12, 26 August 1741.

29 Baugh, *British Naval Administration*, 437.

30 Regarding the infrastructure at Portsmouth during this war, see: PRO, ADM 110/12, 15 August 1739 and 27 June 1740; ADM 110/13, 11 March, 11 November 1743; ADM 110/14, 16 November 1744.

31 PRO, ADM 110/12, 27 March and 4 November 1741.

32 PRO, ADM 110/14, 11 March 1745, 13 May 1746; ADM 110/15, 15 July 1747. See also Baugh, *British Naval Administration*, 437–9.

33 Baugh, *British Naval Administration*, 439.

1749–62: A period of consolidation and expansion

The years 1749–62 were years of both consolidation and expansion. The early years of the 1739–48 war had unfortunately revealed the tendency of suppliers to victual the Royal Navy with poor quality flour, oats and dried peas. Since the inspectors from the Victualling Board found it more difficult to detect this type of fraud in processed rather than unprocessed products, the Victualling Board therefore resolved to buy these products in their natural state and process them itself.[34]

As a result, even more mills and ovens were built in the 1750s. In 1755, for example, new mills were constructed at Redhouse and Dover. The following year, construction at Deptford included an oven built in order to roast peas, and two mills: one for flour and the other for oats, as the oats had to be milled before they were made into oatmeal.[35]

Also, in order to have better control of the beer market, both in terms of quality and so as to rid itself of cartels which were financially detrimental,[36] the Victualling Board developed its own means of production. In 1753 the Commissioners bought a brewery in Gosport to serve Portsmouth's needs, and in 1756 another brewery and cooperage were built, which became operational in the summer of 1757.[37] In Plymouth, the small brewery established at South Down in 1753 had continued to expand. Its production capacity of 80 tuns a week went up to as much as 40 tuns of beer per day in June 1760,[38] that is to say more than 10,000 gallons of beer per day (a tun of beer = 252 gallons). New warehouses were built to stock the beer, ideally brewed before the end of March,[39] and the project for repairs submitted by the master-brewer in 1760 was accepted.[40]

Identical developments took place in the field of meat. The Victualling Board's desire to buy live animals – 'on the hoof' – and to directly undertake the slaughter and processing of meat in order to have better control of the food production process, obliged the Commissioners to continue building slaughterhouses and processing houses, in which the cuts of meat were put into casks and covered in pickling brine to ensure their preservation. Also, storehouses had to be built in which the sealed casks were stocked before

34 Gradish, *The Manning of the British Navy*, 148–9.

35 PRO, ADM 110/18, 26 July 1756, 21 February 1757, 11 April 1757; ADM 110/19, 17 June 1757.

36 Mathias, *The Brewing Industry*, 202.

37 PRO, ADM 110/18, 11 April 1757. See also Mathias, *The Brewing Industry*, 201–2.

38 PRO, ADM 111/50, Letter from the Agent Victualler Ommanney dated 6 June 1760.

39 The provisional costs were estimated at £2,562 (PRO, ADM 110/18, n.d.).

40 PRO, ADM 111/50, 9 April 1760.

Table 4.2 Projected numbers of personnel to be victualled from each port

	Total	London[1]		Portsmouth		Plymouth		Dover		Ireland, West Indies and other Foreign Parts	
		Men	%	Men	%	Men	%	Men	%	Men	%
For 1757[2]	55,000	25,000	45.45	12,000	21.82	10,000	18.18	2,000	3.64	6,000	10.91
For 1758[3]	60,000	27,000	45.00	13,000	21.67	11,000	18.33	2,000	3.33	7,000	11.67
For 1759[4]	60,000	27,000	45.00	13,000	21.67	11,000	18.33	2,000	3.33	7,000	11.67
For 1760[5]	70,000	31,000	44.29	15,000	21.43	13,000	18.57	2,500	3.57	8,500	12.14
For 1761	Here there is a gap in the archives (probably similar figures to the year before)										
For 1762[6]	70,000	31,000	44.29	15,000	21.43	13,000	18.57	2,500	3.57	8,500	12.14
For 1763[7]	70,000	31,000	44.29	15,000	21.43	13,000	18.57	2,500	3.57	8,500	12.14

1 'Port from whence the ships at Deptford, Woolwich, Chatham, Sheerness and the Mediterranean are victualled.'
2 Annual declaration in ADM.111/ 45, 25 October 1756.
3 Annual declaration in ADM.111/ 46, 2 November 1757.
4 Annual declaration in ADM 111/ 48, 17 November 1758.
5 Annual declaration in ADM.111/ 49, 3 October 1759.
6 Annual declaration in ADM.111/ 52, 5 October 1761.
7 Annual declaration in ADM.111/ 54, 15 October 1762.

being loaded onto ships. A new slaughterhouse was thus built in Dover in 1756, and the old one was transformed into a meat-processing house.[41]

Turning to another matter, the water supply system in the victualling yard at Plymouth was significantly improved. In fact, in the spring of 1760 Lord Edgcumbe allowed access to some springs situated on his land near Empacombe in exchange for the right for his cattle to drink from the channels cut in his fields for this purpose, a concession which considerably increased Plymouth's capacity.[42]

The Respective Importance of the Victualling Board Bases in Victualling the Royal Navy

An overall analysis of the number of personnel requiring provisions and of the quantities of provisions involved has to be established before the importance and respective development of each of the naval yards under the jurisdiction of the Victualling Board can be evaluated, in terms of numbers of personnel and facilities. The ADM 111 series with its bookkeeping records relating to the workforce makes such an evaluation possible.

An initial approach would be to examine the annual statement which was submitted in the autumn of each year and which gave the overall number of personnel to be victualled the following year, with a breakdown by port or geographical area. Table 4.2, in which this data is assembled, clearly illustrates the absolute predominance of the port of London. Almost half of all Royal Navy personnel were victualled from London: between 44.29 per cent and 44.45 per cent according to the period. The ports of Chatham and Sheerness do not appear in this breakdown, but are included within London's sphere of activity, and documents also indicate that London was responsible for victualling Mediterranean ports.

The naval personnel to be victualled at Plymouth and Portsmouth together do not equal the numbers of the London personnel. Portsmouth outnumbers Plymouth by 2,000 men, but the gap in relative values narrows between the two ports: the differential of 3.64 per cent at the outbreak of hostilities falls to 3.34 per cent from 1758 and is no greater than 2.86 per cent at the end of 1760. Dover falls far behind, only responsible for victualling some 3.5 per cent of naval personnel. Also of relative importance was the victualling to be carried out 'abroad' (Ireland and overseas): 10.91 per cent to 12.14 per cent, mainly as a result of personnel who had to be victualled in the West Indies, which was the responsibility of private contractors. This topic will be treated

41 PRO, ADM 110/18, 26 July 1756; ADM 110/19, 19 September 1757.

42 PRO, ADM 110/20, 28 April 1760.

Table 4.3 Produce to be sent from London to Portsmouth, Plymouth, Chatham and Dover between 15 May 1757 and 31 December 1757

Product	Unit	Portsmouth	Plymouth	Chatham	Dover	Totals
Cheese	Pound	367,000	200,000 (54.49% of Portsmouth)	27,000 (7.36% of Portsmouth)	30,000 (11.72% of Portsmouth)	624,000
Butter	Pound	140,000	80,000 (57.14% of Portsmouth)	12,000 (8.57% of Portsmouth)	13,000 (9.29% of Portsmouth)	245,000
Olive oil	Gallon	6,000	2,000	/	/	8,000
Vinegar	Gallon	22,000	16,000	500	/	38,500
Beef	Piece	150,000	50,000	6,200	/	206,200
Pork	Piece	225,000	100,000	6,300	/	331,300
Peas	Bushel	13	8,000	900	800	22,700
Groats	Bushel	5,000	2,100	/	500	7,600
Hops	Bag	370	280	/	/	650
Raisins	Pound	65,000	10,000	/	/	75,000
Bread	Bag	14,000	22,000	1,200	/	37,200
Suet	Pound	8,000	10,000	/	/	18,000
Beer	Tun	/	/	200	/	200
Brandy	Gallon	10,000	/	/	/	10,000
Flour	Pound	/	234,000	5,000	28,000	267,000
Oatmeal	Bushel	/	1,000	/	/	1,000
Dates on which deliveries were due		6, 15, 29 June; 13 July; 5, 10, 24 Aug; 9, 19, 21 Sept; 5, 7 Oct; 2, 18, 23 Nov; 9, 19, 21, 23 Dec	6, 15, 29 June; 13, 25 July; 10, 24 Aug; 9, 19, 21 Sept; 7 Oct; 2, 11, 23 Nov; 9, 23 Dec	6, 15 June; 25 July; 10 Aug; 21 Sept; 17 Oct; 2, 23 Nov; 23 Dec	29 June; 10, 24 Aug; 5 Oct; 21, 23 Nov	

Source: Table based on ADM.111/46.

at greater length in chapter 6 concerning the victualling of outlying bases under the jurisdiction of private contractors.

London's role in victualling the Navy was even more important than would appear from this table. In fact, London's central role in dealing with agricultural markets must not be forgotten. London was where the majority of agricultural produce converged, and the Commissioners met in London to agree on victualling policy. Therefore, besides the large number of personnel involved in victualling in the capital, London was also the centre through which nearly all goods required by the other ports under the jurisdiction of the Victualling Board passed.[43] Three or four times a month, the Commissioners ordered exact quantities of produce to be dispatched to these ports.

Table 4.3 details all the produce dispatched from London between 15 May and 31 December 1757. Breaking down this information with reference to the individual ports catered for from London also provides a balance to Table 4.2, which is very hypothetical and does not take into account each port's importance and the activities carried out from those ports. For example, quantities of butter and cheese would appear to be reasonable yardsticks by which to measure a port's level of activity, excluding Gibraltar whence these two items were not dispatched, due to their perishable nature and Gibraltar's remoteness and hotter latitude. Thus, during the second half of 1757, Portsmouth was the second most important port after London; Plymouth lagged far behind, having only received 54.49 per cent and 57.14 per cent of the respective quantities of cheese and butter supplied to Portsmouth. The amount of cheese and butter supplied to Dover only reached 11.72 per cent and 9.29 per cent respectively. As for Chatham – which was included in London's annual personnel statistics – its decline can best be measured in terms of the decline in its goods traffic, as highlighted in the first part of this chapter (7.36 per cent and 8.57 per cent of the quantities of cheese and butter delivered to Portsmouth).

Similarly, in order to interpret future developments, Table 4.4 lists the dispatches of goods from London during the first three quarters of the years 1760[44] and 1761. The increasing importance of Plymouth as far as victualling is concerned is very clear. Compared with the previous table, quantities of butter and cheese received in Plymouth during the three quarters in question in 1760 approached those stocked in Portsmouth during the same period (82.21 per cent against 54.49 per cent for cheese; 75 per cent against 57.14 per cent for butter). However, during 1761, supplies for Plymouth greatly exceeded those for Portsmouth: 138.75 per cent and 135.34 per cent for

43 This aspect will be returned to in the next chapter.

44 The ADM.111 series is incomplete for the period 1 September–31 December 1760.

Table 4.4 Produce sent from London to Portsmouth and Plymouth during the first three quarters of 1760 and 1761

		1760		1761		Totals	
Product	Unit	Portsmouth	Plymouth	Portsmouth	Plymouth	Portsmouth	Plymouth
Butter	Pound	280,000	210,000 (75% of Portsmouth)	240,000	333,000 (138.75% of Portsmouth)	520,000	543,000 (104.42% of Portsmouth)
Olive oil	Gallon	-	-	7,000	3,000	7,000	000
Cheese	Pound	371,000	305,000 (82.21% of Portsmouth)	365,000	494,000 (135.34% of Portsmouth)	736,000	99,000 (108.56% of Portsmouth)
Vinegar	Gallon	28,000	46,000	30,500	33,000	58,500	79,000
Beer	Gallon	-	-	-	-	-	-
Hops	Bag	450	140	330	390	780	530
Flour	Pound	-	550,000	-	530,000	-	1,080,000
Peas	Bushel	13,000	11,000	11,000	14,000	24,000	25,000
Raisins	Pound	70,000	60,000	32,000	31,000	102,000	91,000
Corinthe raisins	Pound	-	-	-	-	-	-
Bread	Bag	-	7,000	-	-	-	7,000
Oatmeal	Bushel	4,000	2,000	2,000	8,800	6,000	10,800
Mustard	Pound	-	1,175	-	-	-	1,175
Molasses	Pound	-	6,930	-	-	-	6,930
Pork	Piece	150,000	100,000	100,000	100,000	250,000	200,000
Beef	Piece	100,000	230,000	100,000	100,000	200,000	330,000
Suet	Pound	8,000	-	-	20,000	8,000	20,000
Groats	Bushel	-	-	2,000	1,000	2,000	1,000

butter and cheese respectively. Plymouth's second position after London, barely perceptible in terms of statistics concerning the number of seamen to be victualled in each port, is, however, confirmed by the higher financial cost for Victualling Board personnel employed at this port, compared to that of Portsmouth, at the end of the first quarter of 1761 (data which has been obtained from the ADM 113 series ledgers), namely, £3,244 7s 6d[45] as against £2,788 10s 10d.[46] This development arises from the geographical situation of Plymouth, the most westerly port of the British Isles. Plymouth thus became the port where vessels destined mainly for America were equipped, and, in particular, it became a supply base, indispensable in the victualling of blockading squadrons operating on the French coast (see Chapter 3), including also, in 1761, the 8,000-strong military expedition to seize Belle-Ile.[47]

Financial records of personnel employed in each of the ports under the jurisdiction of the Victualling Board, held in the archival series ADM 113, offer the possibility of completing this research. It is possible, thanks to these archives, to group personnel together by activity and thus evaluate the respective importance of each of the victualling bases by presenting their profile in terms of infrastructure.

The period chosen for this study was not left to chance. The most obvious choice was to select both a year and the quarter during that year when activity levels were at their greatest, in order to best evaluate the volume and diversity of tasks carried out by those victualling yards. Preliminary research involved taking London as the example – due to its central role it was the benchmark port – and beginning by listing personnel costs at the end of each quarter between 1 January 1756 and 30 September 1763. These results are to be found in Table 4.5, and clearly illustrate that the first quarter of the year corresponds to a period of great activity, followed by the fourth quarter, with the exception of 1758 when the financial burden of the fourth quarter is greater than that of the first three months.

This marked difference between winter and summer months arises from the fact that during the first and fourth quarters *'messers and salters'* were working at full capacity. The *'messers'* (four in number during the whole of the Seven Years War[48]), assisted by *'salters'* (eighteen men during the first quarter of 1761), were responsible for slaughtering the animals, cutting them up, salting the pieces of meat and putting them into casks, and making sure

45 PRO ADM 113/187.

46 PRO ADM 113/225.

47 D. Baugh, *The Global Seven Years War 1754–1763* (Harlow, Longman, 2011), 526–31.

48 In 1756, this concerned Thomas Shephard, John Collingwood, Thomas Gutters and William Toplis (see PRO ADM 113/142); in 1761 this concerned Thomas Shephard, John Collingwood, William Toplis and John Smith (see PRO ADM 113/144).

Table 4.5 Quarterly fluctuations in personnel costs at the London victualling station (1 January 1756–30 September 1763)

Period	Cost	Source	Period	Cost	Source
January–March 1756	£6,002/0/8	ADM 113/142	January–March 1760	£7,430/17/5	ADM 113/144
April–June 1756	£4,472/1/2	“	April–June 1760	£6,052/11/11	“
July–Sept. 1756	£4,089/5/3	“	July–Sept. 1760	£5,391/4/0	“
October–Dec. 1756	£188/14/5	“	October–Dec. 1760	£7,135/1/11	“
January–March 1757	£6,056/14/6	“	January–March 1761	£8,467/12/05	ADM 113/145
April–June 1757	£4,679/0/5	“	April–June 1761	£5,460/15/11	“
July–Sept. 1757	£4,878/11/9	“	July–Sept. 1761	£4,967/9/6	“
October–Dec. 1757	£5,722/8/12	“	October–Dec. 1761	£7,088/15/9	“
January–March 1758	£6,209/14/8	ADM 113/143	January–March 1762	£8,030/2/1	ADM 113/146
April–June 1758	£4,873/2/0½	“	April–June 1762	£5,561/13/0	“
July–Sept. 1758	£4,953/15/6	“	July–Sept. 1762	£5,193/13/6	“
October–Dec. 1758	£7,062/5/1	“	October–Dec. 1762	£6,392/15/7	“
January–March 1759	£7,702/0/7	“	January–March 1763	£3,058/15/11	ADM 113/147
April–June 1759	£5,672/11/4	“	April–June 1763	£2,195/-/5	“
July–Sept. 1759	£5,113/15/11	“	July–Sept. 1763	£2,049/14/9	“
October–Dec. 1759	£6,756/10/9	“			

their work procedures conformed strictly to the rules drawn up by the Victualling Board.[49] They were not permanent employees but worked only from the autumn, when the seasonal migration of livestock had ended[50] and the Commissioners purchased animals, until the spring, when a rise in tempera-

49 Instructions for Samuel Rowe, employee in the cutting up shop in the port of London (27 June 1716), in Baugh, *British Naval Administration*, 413–14.

50 The livestock came mainly from two regions: Wales and Scotland. Welsh livestock crossed the Midlands, Essex and Kent. Scottish livestock travelled across the rich pastures of Norfolk and Suffolk. These routes have been studied by G.E. Fussel, C. Goodman and C. Skeel (Fussel and Goodman, 'Eighteenth-century traffic in live-stock', *Economic History*, February 1936, 214–36; Skeel, 'The cattle trade between Wales and England from the

tures could have a harmful effect on meat processing. In 1761 and 1762 the activities of the '*messers*' and '*salters*' ran into the second quarter, but the modest sums released as payment at the end of these two quarters indicate their limited nature. Each '*messer*' received £2 6s 8d per full day's work and the '*salters*' received 2 shillings each, to which was added an indemnity which was calculated in proportion to the amount of work carried out, namely, 12d per hundredweight which the *messers* and *salters* had to divide up amongst themselves. The *messers* also benefited collectively from an additional 2d per oxen.[51]

Thus the first quarter of the year 1761 has been used as this chapter's period of study. The quarter in question corresponds to the highest level of expenditure spent on the workforce during the whole of the Seven Years War. In total, for the five home yards under the jurisdiction of the Victualling Board, this expenditure reached £15,856 for 1,333 personnel (see Table 4.6). London employed nearly 44 per cent of these personnel; Plymouth exceeded Portsmouth with 353 employees as against 248 for Portsmouth; these three bases themselves totalled 1,205 employees, that is to say almost 90 per cent (88.6 per cent) of the workforce.

Table 4.6 Number of employees and personnel costs at victualling yards at the end of the first quarter of 1761

Victualling yard	Cost	Enlistment on 1 January	The yard's enlistment as a percentage of total enlistment
London	£8,467	604	44.41
Plymouth	£3,244	353	25.96
Portsmouth	£2,788	248	18.24
Dover	£1,022	69	5.07
Chatham	£335	59	4.33
Gibraltar	£2,116	27	1.99
Total	£17,972[1]	1,360	100

1 Twenty-four interim workers reinforced the workforce of twenty-seven permanent workers during the first quarter.

Table 4.7 groups together all the figures relating to the port of London by sector of activity: in this way, their importance can be evaluated. This yard was the only one to have the complete range of facilities necessary to victual a whole fleet. The yard had a cooperage (168 men), two flour-milling works, one at Rotherhithe (27), the other at Redhouse (15), two bakeries, one at

fifteenth to the nineteenth centuries', *Transactions of the Royal Historical Society*, 4th series, IX (1926), 135–98).

51 PRO ADM 113/144. Accounts register from the Victualling Board's base in London during the first quarter of 1761.

Table 4.7 Breakdown of the London victualling base workforce by sector of activity (1 January 1761)

I – Personnel in Payment (excluding office staff – see Appendix 1)	
Cooperage: 168	
92 coopers	
5 servants to the master butcher and to the master cooper	
6 other servants	
65 labourers	
Brewhouse: 51	
1 master brewer	
2 coopers	
2 stockers	
2 millers	
6 tunmen	
1 foreman	
1 cask prover	
36 yardmen	
Mill: 42	
At Rotherhithe	
4 millers	
1 pastryman	
1 millwright	
1 measurer of bran and small coal	
1 bread weigher	
1 foreman	
17 labourers	
1 occasional pondsman	
At Redhouse	
5 millers	
1 pastryman 1 extra hand	x 4 teams
1 inspector of corn	
1 millwright	
Bakehouses: 130	
At Rotherhithe	
1 furner 1 mate 1 driver 2 boardmen	per oven (x 7 ovens)
1 pastryman	
At Redhouse	
1 furner 1 mate 1 driver 2 boardmen	per oven (x 7 ovens)
1 foreman	
33 bakehouse labourers	

Warehouses and Factories: 116
Flour loft
4 labourers
Oat mill
1 foreman
1 oatmaker
5 labourers
Cooperage loft
1 foreman cooper
1 cooper
7 labourers
Bread loft
1 foreman
2 bread weighers
16 labourers
Various employees of the warehouses and factories
1 exterminator of vermin
4 bag menders
25 gatekeepers and messengers
1 chimney sweep
40 wharf labourers
Cuttinghouse: 55
33 pickle yard's labourers
4 messers
18 salters
Loading Operations: 42
For the sloop *Princess Augusta*
1 master
1 servant
1 mate
4 able seamen
For the sloop *Caesar*
1 master
1 servant
1 mate
3 able seamen
29 wharf labourers
Total number of employees: 604
II – Specific Contracts
Cartage: George Wheeler
Waterage and petty freight: Charles Harwood
Provision of horses to grind oatmeal: Richard Haws
For the record:
Persons responsible for slaughtering the animals and salting the meat (see Cuttinghouse)
Chimney sweeping: Suzanne Mills (see Warehouses)
Exterminator of vermin: the Crossley child (see Warehouses)

Source: Table drawn up from data obtained from ADM 113/144.

Table 4.8 Breakdown of the Portsmouth yard workforce, sector by sector (1 January 1761)

I – Personnel in Payment	
Officers and their Staff: 17	
1 Agent (Matthew Oakes)	
1 storekeeper	
1 clerk of the checque	
1 master cooper	
5 clerks to the agent	
3 clerks to the storekeeper	
3 clerks to the checque	
2 clerks to the master cooper	
Mill: 8	
1 master miller	
2 millers	
4 pastrymen	
1 servant to the master miller	
Bakehouse: 21	
1 foreman	
1 furner	per team (x 5 teams)
1 mate	per team (x 5 teams)
1 driver	per team (x 5 teams)
1 idleman	per team (x 5 teams)
Warehouses and Factories: 82	
12 labourers	
62 extra labourers	
7 coopers	
1 porter	
Loading Operations: 3	
1 superintendent	
1 hoyman	
Cooperage: 53	
Lord James Robinson	
1 servant to J. Robinson	
1 foreman	
1 other servant	
4 servants to the master cooper	
1 servant to the foreman	
44 coopers	

Brewhouse: 64
1 master brewer
1 clerk
2 stockers
2 millers
2 tun men
52 labourers
1 watchman
3 coopers
Total number of workers: 248
II – Special Contracts
Inspectors and foremen of the production chain
for the inspection of beef: William Hillyar
for the inspection of sheep: John Bridger and Thomas Randall
for the inspection of pigs: William Hillyar

Source: Table drawn up from data in ADM 113/225.

Rotherhithe (36), the other at Redhouse (61), a brewery at Hartshorn (51), a cuttinghouse (55) and several storehouses; a list of their employees can be seen in Table 4.7. In addition, there was another centre of activity, consisting of loading, handling and transport operations, which required forty-two employees; and the crews from two sloops belonging to the Victualling Board, the *Princess Augusta* (seven men) and the *Caesar* (six), and, last but not least, the twenty-nine packers assigned to the wharf.

Without going through all the information in Table 4.7, it is interesting to highlight that the yard employed twenty-five gatekeepers and messengers, four bagmenders, one person responsible for killing vermin ('the Crossley child', paid £15 per year), and a chimney-sweep, Suzanne Mills, employed to clean the yard's 143 chimneys at the rate of 6d per chimney. For the sake of convenience, all these elements have been grouped with the personnel assigned to the stores. In addition, a contract was also drawn up with a Richard Haws to hire horses for the flour-milling works at Redhouse, so that the oats could be milled at the rate of 14 shillings per day. Appendix 7 breaks down employee costs according to their activity.

Apart from the absence of a slaughterhouse ('cuttinghouse'), Portsmouth had all the facilities and infrastructure necessary to carry out the different activities (see Table 4.8). Consequently, for meat, they relied on victualling carried out from London. They were allowed to buy live animals, which were loaded onto the ships destined for long voyages,[52] which permitted them to have a small amount of fresh meat available in the course of a campaign.

52 See Chapter 3.

Table 4.9 Breakdown of workers in the Plymouth yard, sector by sector (1 January 1761)

I – Personnel in Payment	
Officers and their Staff: 13	
1 Agent (John Ommaney) 1 storekeeper 1 checque 1 master cooper 4 clerks to the agent 3 clerks to the storekeeper 2 clerks to the cheque	
Bakehouse: 42	
1 foreman 1 foreman's servant	
1 furner 1 mate 1 driver 1 boardman 1 idleman	} per oven (x 8 ovens)
Warehouses: 130	
108 labourers 1 gatekeeper 1 bagmender 18 watchmen 1 foreman cheesemonger 1 person for cleaning the water pipes, and looking after the air cocks and the hatches	
Loading Operations: 1	
1 boatswain of the wharfers	
Cooperage: 59	
At Plymouth: 17 coopers 3 servants to the master cooper 1 servant to the foreman 1 servant 1 foreman **At South Down:** 1 foreman 2 servants to the foreman 33 coopers	
Brewhouse at South Down: 108	
1 master brewer 1 clerk 1 checque 1 servant 46 brewers	
Total number of employees: 353	

II – Special Contracts	
The charter of the following ships (for victualling, manning and navigating)	
Catwater sloop: John Horlock	*Mary sloop* (≈ 36 tons): Joseph Brent
Mary (≈ 30 tons): Georges Marshall	*Jenny* (≈ 45 tons): Thomas Wrench
Betsey (≈ 30 tons): Thomas Wrench	*Sarah sloop* (≈ 70 tons): Joseph Brent
Carrying oxen and sheep to 'crimmel' (Cremel) passage: John Julian (1,035 oxen, 1,298 sheep)	
Surveying oxen, sheep and hogs: Samuel Stephens and John Adams	
Purchasing of animals on the hoof (oxen, sheep and hogs): Samuel Stephens and Philip Smith	

Source: Table drawn up from data in ADM 113/187.

Contracts to this effect were drawn up with private contractors, with inspection of the animals before embarkation. As far as the loading operations of ships were concerned, they were carried out under the supervision of a superintendent, who also had to inspect the holds of the chartered ships, assisted by two hoymen.

Plymouth did not seem to have a slaughterhouse either,[53] as meat was sent from London. In order to be able to meet future demands for live animals made by the squadron commanded by Sir Edward Hawke, which was blockading the French coast, contracts were drawn up by Agent John Ommanney for the purchase, forwarding and inspection of the required animals: in this way, 1,035 oxen and 1,298 sheep appear to have been driven to Plymouth during the first three months of 1761.

Unlike Portsmouth, Plymouth did not have a flour-milling works. This fact is corroborated in Tables 4.3 and 4.4 which list all the produce sent from London to the yards directly administered by the Victualling Board. In fact, while no flour was sent to Portsmouth, 224,000 pounds of flour were dispatched to Plymouth between 1 June and 31 December 1757, and 1,080,000 pounds of flour during the quarters considered in 1760 and 1761.

Therefore, by 1761, even though Plymouth had become the second most important victualling port after London (see above), it still did not have the extensive facilities enjoyed by Portsmouth. The large number of storekeepers assigned to the warehouses (130 in Plymouth, as against 82 in Portsmouth) shows the specific character of Plymouth's development. Without a slaughterhouse or flour-milling works, Plymouth depended on London a great deal more than Portsmouth did, and attempted to counterbalance its evident lack

53 More or less in use, since, in the documents consulted, a small slaughterhouse is mentioned which would have been built in 1756 (National Maritime Museum, ADM-C.541, 17 and 26 February 1756). A proposition was made in September 1757 for its enlargement (PRO, ADM 110/18, 19 September 1757). The estimate for enlargement was £171 8d.

of manufacturing facilities with a high number of storehouses. Its cooperage only employed twenty-three people, whereas in Portsmouth there were fifty-three employees.

Six sloops chartered as victualling vessels (see Table 4.9) for the western squadron blockading the French coast illustrate the extent to which the presence of this powerful squadron played a role in Plymouth's rapid development. The only point on which Plymouth exceeded Portsmouth was in terms of its bakery, which was composed of eight ovens as opposed to four in Portsmouth, and where forty-two employees worked, as opposed to twenty-one in Portsmouth. In London, Plymouth and Dover, five men were assigned to each oven, as opposed to only four in Portsmouth.

The last three victualling yards under the auspices of the Victualling Board (Dover, Chatham and Gibraltar) were fairly modest compared with London, Plymouth and Portsmouth. The largest number of personnel was in Dover (sixty-nine men), and these employees only accounted for 5.07 per cent of the 1,360 Victualling Board employees distributed throughout the six bases. Dover did not have a flour-milling works or a brewery, but it did have a slaughterhouse in which seventeen people worked under the supervision of a messer called James Tolputt. This activity was already apparent during the second half of 1757 for, as Table 4.3 illustrates, Dover was the only port not to be victualled by London in terms of pork and beef, which is evidence of its self-sufficiency in terms of meat. During the first quarter of 1761, 683 oxen and 500 pigs were slaughtered and processed.[54]

The bakery in Dover seems to have been equally important for the yard since it had five ovens, that is to say as many as the bakery in Portsmouth: twenty people were employed there during the first quarter of 1761, compared with twenty-one people in Portsmouth.

Finally, in Table 4.10, information relating to Dover has been grouped together. It is worth noting that, just as in London, four private individuals (Phineas Stringer, Martha Atherden, John Worthington and Edward Atherden) were under contract for haulage, which had to be carried out between the upper stores, probably situated on the cliffs, and the lower stores, situated in the port. The rate for each trip was 2s/6d for bread, 1s/4d for beer, and 8d for a load of empty casks.[55]

In 1761, Chatham only played a minor role in the victualling of the Navy. Chatham did not have flour-milling works, a brewery or a slaughterhouse, and one cannot truly talk of a cooperage or of a bakery for premises in which only two and three men were employed respectively. In fact, Chatham, as

54 PRO, ADM 113/90.

55 PRO, ADM 113/90.

Table 4.10 Breakdown of workers at the Dover yard, sector by sector (1 January 1761)

I – Personnel in Payment	
Officers and their Staff: 8	
1 Agent (Michael Russel)	
1 master storekeeper	
1 checque	
1 master cooper	
2 clerks to the agent	
1 clerk to the storekeeper	
1 foreman to the bakehouse	
Cooperage: 11	
10 coopers	
1 apprentice	
Bakehouse: 20	
1 furner }	
1 mate }	
1 baker }	per oven (x 4 ovens)
1 boardman }	
1 boy }	
Warehouses: 30	
30 labourers	
Total number of employees: 69	
II – Special Contracts	
Cartage:	
Phineas Stringer	
Martha Atherden	
Edw. Atherden	
John Worthington	
1 messer: James Tolputt	
14 salters	
2 randers	

Source: Table drawn up from data in ADM 113/90.

the earlier part of this chapter illustrated, was little more than a warehouse, dependent on London, and was only maintained because it was well situated and well sheltered at the mouth of the Thames and near the Nore anchorage. The fact that personnel at Chatham were responsible for handling on the two small loading ships, a skiff and a hoy, is significant in this respect.

As for the victualling yard in Gibraltar, only twenty-seven employees worked there full-time during the first quarter of 1761. They were divided into two areas of activity: the coopers and the storekeepers, helped by some twenty-four temporary labourers. Gibraltar, like Chatham, was little more

Table 4.11 Breakdown of workers at the Chatham yard, sector by sector (1 January 1761)

I – Personnel in Payment
Officers and their Staff: 5
1 Agent (Henry Hutchinson) 1 master storekeeper 2 clerks to the agent 1 clerk to the storekeeper
Cooperage: 3
1 master cooper 1 cooper 1 servant to the master cooper
Miscellaneous Personnel: 44
26 labourers 12 night watchmen 4 warders 2 bakers
Crew of Small Boats Required for Loading Purposes: 7
Crew of Chatham's boy 1 master 1 mate 1 able seaman 1 master's servant ***Crew of Chatham's yaul*** 1 master 1 mate 1 master's servant
Total number of personnel: 59
II – Special Contracts
None

Source: Table drawn up from data in ADM 113/13.

Table 4.12 Breakdown of workers in the Gibraltar yard, sector by sector (1 January 1761)

I – Personnel in Payment
1 Agent (John Lewis) 2 clerks 1 store clerk 8 coopers 8 labourers 1 watchman 6 crewmen
Total: 27
II – Special Contracts
None

Source: Table drawn up from data in ADM 113/109.

than a distribution centre, victualled by London and by local contracts for products that could be described as 'finished products', not requiring any further processing.

Table 4.13, in which the main activities of each victualling yard have been grouped, summarises the information presented in this chapter.

Table 4.13 Summary of personnel, sector by sector, in the victualling yards under the jurisdiction of the Victualling Board

	London	Plymouth	Portsmouth	Dover	Chatham	Gibraltar
Cooperage	168	59	53	11	3	8
Milling	42	/	8	/	/	/
Bakehouse	130 (19 ovens)	42 (8 ovens)	21 (5 ovens)	20 (4 ovens)	1	/
Brewery	51	108	64	/	/	/
Slaughter-house	55	/	/	17	/	/
Warehouses	116	130	82	30	26	8

Having examined the evolution of the Victualling Board's bases from a geo-strategic perspective, the next chapter will provide a detailed analysis of the day-to-day running of one of the yards. Plymouth was chosen for this example, since it had the greatest victualling activity – after London – of all the bases during the final years of the Seven Years War.

5

An Example of a Victualling Yard under the Jurisdiction of the Victualling Board: Plymouth, Satellite of British Logistical Power

This chapter will approach its subject matter from a different slant from that taken in the previous chapters of this book. Up to now, the aim of this volume has been to highlight a system that was progressively developed by the Victualling Board in order to victual the Royal Navy, and as a consequence the approach taken has been that of presenting this as a whole, in a macro-economic fashion, starting with an examination of the work of the central Commissioners at top level. This chapter will complete the study by taking a micro-economic approach and analysing how victualling was organised at the local level. This will be done by undertaking a thorough investigation of the victualling yard at Plymouth. Of all the major ports, Plymouth was considered to be the most suitable for this study, for a variety of reasons, including its role in victualling an ever-increasing number of vessels despite its incomplete infrastructure (see Chapter 4), a task that was not without certain challenges.

The first part of this chapter will analyse the division of tasks among the men in charge of victualling at Plymouth, and the second section will then examine the mode of management and the different sources for this west-ernmost of home victualling yards, working in the closest proximity to naval operations.

Plymouth Personnel and Victualling

The Agent Victualler

In charge of each victualling yard under the jurisdiction of the Victualling Board – excluding London which was directly managed by the Commissioners – was an Agent Victualler. This position had been held by John Ommanney in Plymouth since 12 December 1754.[1] Consequently, when the

1 PRO, ADM 110/20, 2 March 1761.

war broke out, the Commissioners found themselves in the enviable position of having an experienced, 53-year-old representative in Plymouth.[2] A study of Mr Ommanney's correspondence with the Commissioners throughout the war reveals a man who took his work seriously, showed great responsibility and deservedly gained the respect and confidence of his hierarchical superiors. Consequently, the atmosphere of mutual trust and confidence that reigned between the Plymouth Agent and the managers at the Victualling Board guaranteed that the job at hand was done quickly and efficiently.

In March 1761, the Commissioners wrote a letter to accompany the list that had been compiled of all the officers at each port. It is clear from this list, which noted the ages of the officers, that the Commissioners could indeed congratulate themselves on employing men with solid work experience. The list and letter are reproduced in Appendix 8.

Similarly to his counterparts in Portsmouth, Dover, Chatham and Gibraltar, John Ommanney managed the whole victualling complex in Plymouth. He took delivery of goods arriving from London or any other production site. He ensured that the goods were painstakingly stored and made certain that all missing or rotten foodstuffs were replaced. He was in charge of food processing, and supervised and adapted the brewery's output, and the production of bread and biscuits, according to demand. He was then responsible for loading the victuals onto the vessels, paying careful attention to quantities, which were calculated according to the length of the voyage. The victuals were either loaded onto war ships or onto victualling vessels, which were frequently chartered in order to increase tonnage space, and it was up to the agent to manage the chartering and to decide how many vessels were required. He had to inform the Commissioners of where the vessel(s) had been chartered from,[3] the charter fee (usually between 11 and 14 shillings per ton per month) and the conditions of hire, and last but not least, at any given time he had to know the exact location of the ships.[4] He was also responsible for arranging the victualling ships' escorts, since they always sailed in a convoy.

2 Ommanney had been a junior officer at Portsmouth in 1743. From there he had moved to become Storekeeper at Woolwich by 1748, when he was 'most strongly recommended' for preferment by his patron the Duke of Newcastle. The First Lord of the Admiralty, the Duke of Bedford, then suggested he be advanced to Clerk of the Survey at Portsmouth (Baugh, *British Naval Administration*, 33–4, 484).

3 PRO, ADM 111/51, Letter from the Navy Board dated 23 April 1761.

4 Letters from the Commissioners of the Victualling Board to their agents in Portsmouth and Plymouth asking them to inform them of the return of ships from the fleet, and when they could set sail again. PRO, ADM 111/51, 25 April 1761.

In fact, he could not allow a ship to set sail without a convoy unless he had special permission from the Commissioners.[5]

The Agent was also responsible for the financial management of the yard's various activities. Just like his counterparts in Portsmouth, Dover and Chatham, he received funds twice a year (on 25 March and 20 September), to cover the yard's financial needs. Table 5.1 indicates the sums of money allocated to each of the four main victualling yards over the period of one year – that is, for all activities which took place between 21 September 1760 and 25 March 1761, and then from 25 March 1761 to 21 September 1761. The table shows that the allocated sums correspond to the last penny to the accounts registers for each port which were drawn up quarterly, and which were the object of study in Chapter 4 (cf. the ADM 113 series).

Table 5.1 Victualling funds apportioned to the Agents of the four main ports on 20 September 1760 and 25 March 1761[6]

Port	Name of victualling board agent	Payment received on 20 September 1760	Payment received on 25 March 1761
Plymouth	John Ommanney	£6,143/2/0	£6,552/14/10
Portsmouth	Matthew Oakes	£4,638/14/7	£5,558/17/7
Dover	Michael Russell	£1,626/0/2	£1,937/18/7
Chatham	Henry Hutchinson	£736/18/3	£697/19/10

Source: PRO, ADM 111/52 (7 August 1761); ADM 111/ 53 (27 January 1762).

Besides these habitual funds, the Agent could also request an advance – called an imprest – for contingency measures due to the war. This money would be used in paying workers hired on a daily basis who were taken on in moments of exceptional need – for example, when there were special contracts, or when vessels, or indeed the base itself, required immediate repairs. By their very nature, the dates and amounts of these imprests were variable. However, in the period between September and November 1761, the sum of £3,000 was requested. On each occasion the Agent was reminded that he had to forward the accounts register for all expenditures, and could not draw up invoices without the Commissioners' approval.[7]

For managing and coordinating all these tasks, John Ommanney and his counterparts in the other ports, including Gibraltar, received an annual salary

5 PRO, ADM 110/20, 2 April 1760.

6 PRO, ADM 111/52, 7 August 1761; ADM 111/53, 27 January 1762.

7 PRO, ADM 111/50, 30 July 1760.

of £200.[8] The number of office clerks attached to each agent varied according to the port. In Plymouth, Mr Ommanney had four clerks, whereas Portsmouth had five, but Chatham, Gibraltar and Dover only had two. The clerks were paid about £40 per annum.[9]

The Agent's subordinate officers

In order to help the Agent fulfil his duties, the Plymouth yard employed five subordinate officers, who each had a very specific task, for an annual salary of £60. These five men had the following roles: storekeeper, clerk of the cheque, master cooper, foreman baker and master brewer. Each position will be described in some detail below.

The *storekeeper* maintained and managed the warehouses. He had three assistants, and there were a further 130 employees under the assistants' iron rule.[10] The storekeeper was required by the Commissioners to be very strict towards his subordinates and not to hesitate in laying them off for the slightest transgression.[11]

A certain Mr Forest held this position until his death in March 1760. Then, John Watkins, who had been in the service for a year, was the Commissioners' preferred choice. According to them, he had all the attributes required for the position and was consequently appointed by the Lords of the Admiralty on 12 March 1760. Upon his death in October 1761, a Mr Clodd was appointed, but on 21 October 1761 the storekeeper in Portsmouth also died, and consequently Mr Clodd was sent off to Hampshire to replace him. Three candidates were then presented in Plymouth to replace Clodd: Messrs Mayor and Clyde, both office clerks to the warehouses' accountant, and a Mr Bunce, first employed by Ommanney. It was the latter candidate who was finally appointed. It is clear that experience and a sense of professionalism were the leading factors in these career nominations.[12]

The *clerk of the cheque*, whose hierarchical superior was the Agent, was responsible for all the port's finances. In July 1759, Alexander Gordon took over this post from a Mr John Peury. Both in Portsmouth and in Plymouth the clerk of the cheque had two office clerks working under him. Working hours for all these employees (assistants and officers) were from six o'clock in the morning until eight o'clock at night, with a one-hour pause for lunch and a two-hour break for dinner.

8 ADM 113/90, 113/109, 113/144, 113/187 and 113/225.

9 ADM 113/90, 113/109, 113/144, 113/187 and 113/225.

10 In the first quarter of 1761. See Table 4.9 in Chapter 4.

11 PRO, ADM 111/50, 30 July 1760.

12 PRO, ADM 111/50, 12 March 1760; ADM 111/52, 2 and 21 October 1761.

The *foreman baker* supervised the manufacture of bread and biscuits. He had forty-one employees working under him, eight of them at the ovens. The numbers of batches per day were often modified to account for precise needs; consequently fluctuations in bread and biscuit production were the norm. It appears that about sixteen batches per day were considered to be average production, but in April 1759 the demand was such that output increased to twenty batches per day. In the space of six months, Plymouth had to send 42,370 sacks of bread to the Americas alone. Since the bakery could not produce sufficient quantities (even at the rate of twenty batches per day), contracts were drawn up with the town's bakers to satisfy the demand for biscuits. It is clear that even with the best of intentions and with the workers being pushed to their physical limits, the yard's bakery was unable to meet the requirements. The Commissioners voiced their apprehension to the Lords of the Admiralty when the workers began to express their dissatisfaction, complaining about their working conditions and, in some cases, going on strike. They demanded a wage increase of 2½d per batch. It should be remembered that if the foreman baker earned 23 shillings and 3d per day, his assistants and apprentices only earned 2d per batch – a rate which had not been increased in over forty years. But the Commissioners were fearful that an increase in wages would soon be demanded in the bakeries of other ports, and fiercely opposed the increase, and simply presented the Lords of the Admiralty with a list of all the 'rebels' who had refused to work.[13]

Unfortunately, the outcome of this crisis was not recorded. All that is known is that a few months later the Commissioners did their utmost to arouse interest in the trade. In fact, at the beginning of 1760 the Commissioners wanted to take on a baker and three apprentices in Plymouth. Consequently, in April of that year, Daniel Collins was hired as baker and was provided with 25 shillings to help him move from London to Plymouth. On 30 May, a Mr Andrew Butterfly also proposed his services and also received a travel allowance of 25 shillings from John Everall, the Commissioner's agent who was responsible for minor financial transactions.[14]

In fact, it was most probably the marked drop in demand for bread and biscuits from the beginning of 1760 that allowed the crisis to settle. By February 1760 the number of batches required per day had once again decreased to sixteen, and it even reached an all-time low of fourteen batches in July 1760.[15] The last fortnight in January 1762 again saw an increase in demand, but it did not reach the heights of the spring of 1759. In fact by

13 PRO, ADM 110/19, 27 April 1759.

14 PRO, ADM 110/20, 2 March 1761.

15 PRO, ADM 111/50, 1 February and 21 July 1760.

December 1761, ever-increasing confidence that an end to hostilities was not far away led to the closure of some ovens.

The *master cooper* oversaw the manufacture and repair of barrels, casks and drums. Richard Pett held this position from 27 October 1751 and throughout the war. In March 1761, when Pett turned 67, he became the oldest officer in the service of the Victualling Board in all its ports.[16] At the time he was responsible for twenty-three employees in the small cooperage in Plymouth. The previous year, in 1760, one of his employees had been transferred to London in order to inspect casks of meat coming from Ireland.[17]

The ADM 113 series indicates that in all ports the cooperage workers were divided into three groups: there were those who worked on the wood (*block work*); those who worked on the hoops (*hooping work*); and those whose task it was to seal the casks or open them in order to add preservatives to the meat inside (*heading and unheading casks*).[18]

From 1 October 1756 the *master brewer* was Edward Williams.[19] With John Ommanney acting as middleman, Williams's task was to buy malt locally and then to oversee all the steps leading to the production of beer. Unlike the other four subordinate officers, the master brewer could voice his concerns directly to the Commissioners. Hence, when the main vat broke down and was considered to be irreparable, Williams was in a position to contact the London Office immediately so that repairs could begin as soon as possible. A boilermaker called William Johnson was eventually asked to make a new vat, with the help of other boilermakers sent especially from London. Nonetheless, records show that between the date when the Commissioners learned that the vat was no longer functioning, and the date when the London boilermakers were sent to Plymouth, four months had passed.[20]

16 PRO, ADM 110/20, 2 March 1761.

17 PRO, ADM 111/50, 7 and 16 January; 18 February 1760.

18 PRO, ADM 113/90, 113/109, 113/144, 113/187 and 113/225.

19 PRO, ADM 110/20, 2 March 1761.

20 PRO, ADM 111/50, 11 April and 29 August 1760. The brewing of beer at Plymouth was the one significant cause of dissatisfaction as regards its victualling of the fleet. Complaints by naval officers of 'sour and stinking beer' caused the dismissal of Williams's predecessor, Richard Herring. Further complaints by Admiral Hawke in the crucial summer of 1759 caused the Admiralty to order that a Victualling Board Commissioner, Robert Pett, be sent down to investigate. He reported that the master brewer was fully qualified and acted according to all the standard procedures. The Board blamed that summer's 'uncommon hot weather' which made it difficult to control fermentation. Gradish posits that the faulty vat might have been the problem (Gradish, *Manning of the British Navy*, 157–8).

A time of tension and of temporary hired labour

The victualling process in time of war had to operate at optimal efficiency, since a delay in just one sector would have a negative impact on the whole operation. Nowhere was this more of a problem than at the level of manpower, especially concerning the loading and unloading of cargo. In times of shortage of handling personnel, John Ommanney would simply take on militiamen as temporary hired labour.

In January 1760, even though Ommanney resorted to this solution and took on militiamen,[21] the atmosphere of tension nonetheless grew, in the knowledge that Sir Edward Hawke's fleet, which was patrolling Quiberon Bay, needed to be victualled. By mid-January, the Admiralty deemed it necessary to send some men from London to help with the loading of provisions.[22] On 18 February, Ommanney hired a further thirty militiamen.[23] It should be noted, however, that the Victualling Board's Agent in Plymouth also showed great initiative in attempting to reduce any delays in his yard by hiring temporary personnel without permission – a deed for which he would later be congratulated. In July 1761 he again recruited temporary labour in order to satisfy Commodore Keppel's fleet based at Belle-Ile. Once again, this temporary workforce was hired in order to load the vessels.[24] Finally, in February 1762, the Commissioners requested Ommanney to hire some more hands and operate a round-the-clock service in loading the vessels. This specific need for manpower can be linked to the substantial demand for bread that occurred in January 1762.

Finally, these examples also illustrate the extent to which the Victualling Board could count on credit terms, a situation which had changed considerably from that of past wars and which showed that they now benefited from a certain financial breathing space.

Resources and Management of Victualling at Plymouth

Every year the victualling station at Plymouth received considerable quantities of foodstuffs that were already processed or ready to be processed at the yard. Most of these victuals were sent from London, but some were the result of particular contracts drawn up for Plymouth. In addition, provisions were sometimes sent, as a last resort, from other ports.

21 PRO, ADM 111/50, 7 January 1760.

22 PRO, ADM 111/50, 16 January 1760.

23 PRO, ADM 111/50, 18 February 1760.

24 PRO, ADM 111/50, 1 July 1761.

Victuals sent from London

As was seen in Chapter 3, the Victualling Board Commissioners were responsible for sending the provisions to Plymouth as well as to the other ports.

Throughout the war, regular dispatching was suspended. However, it should be noted that when the Victualling Board ordered John Ommanney to send large quantities of victuals to the western squadron, he rushed to the task, since food shortages had to be avoided at all costs. Table 5.2 shows the method that was adopted in this particular operation. As soon as Ommanney received orders from the Commissioners on 30 January 1762 to send provisions immediately to Sir Thomas Stanhope at Basque Road and Captain Mann at Belle-Ile, the depleted Plymouth warehouses were re-stocked two days later.

Table 5.2 Provisions sent to Plymouth in order to re-stock the warehouses

		Sent from Plymouth on 30/01/62			
Produce	Units	To Basque Road	To Belle-Ile	Total	Sent to Plymouth on 01/02/62
Bread	bag	2,863	1,644	4,507	4,507
Pork	piece	21,665	17,860	39,525	39,525
Suet	pound	9,879	6,750	16,629	16,629
Flour	pound	-	91,035	91,035	91,035
Corinth raisins	pound	9,878	6,750	16,628	16,629
Peas	bushel	1,228	735	1,963	1,963
Oatmeal porridge	bushel	1,706	1,128	2,834	2,834
Butter	pound	4,367	13,669	18,036	18,036
Cheese	pound	9,188	19,719	28,907	28,907
Vinegar	gallon	3,932	-	3,932	3,932
Coal	cauldron	20	56	76	-
Candles	pound	3,400	3,746	7,146	-

Source: PRO, ADM 111/53 (30 January and 1 February 1762).

It appears, therefore, that during the war years, the provisioning of yards took place according to the fluctuating needs of the troops and fleets, and not according to a fixed calendar.

Victuals resulting from contracts signed on Plymouth's behalf

There were basically three types of foodstuff in this category: wine, malt and meat. Apart from the casks of beef and pork, which were directly sent from London, the victualling of Plymouth in meat supplies was the result of the establishment of specific contracts. Just as was the case in London, Portsmouth and Dover, a private contractor supplied Plymouth with fresh meat. The contracts were drawn up on a quarterly basis, and whilst they did not require the supplier to provide a fixed quantity of the product, he was nonetheless bound to satisfy whatever demands were made for fresh meat.

A document which was drawn up at Parliament's request illustrates the importance of this type of victualling in the period from 1 January 1759 to 21 November 1761, for the ports of London, Portsmouth and Plymouth.

Table 5.3 Fresh meat supplied to the ports of London, Plymouth and Portsmouth between 1 January 1759 and 21 November 1761

Port	1759	1760	1761[1]
London	760 cwt[2]	1,761 cwt	1,025 cwt
Portsmouth	27,628 cwt	32,843 cwt	23,709 cwt
Plymouth	20,005 cwt (of which 5,661 cwt was for the fleet operating off the French coast)	30,555 cwt (of which 6,266 cwt was for the fleet operating off the French coast)	24,213 cwt (of which 9,405 cwt was for the fleet operating off the French coast)

1 Until 20 November.
2 Hundredweight.

Source: ADM.110 19, ff. 396–9.

Table 5.3 shows that if Portsmouth in 1759 and 1760 was supplied with the most meat, growth of demand in Plymouth was proportionally greater. This was due to the fact that Plymouth victualled the fleet operating in waters off the French coast, and it thus surpassed Portsmouth in fresh meat victualling during the first eleven months of 1761. London's much smaller demand was a reflection of the dichotomy between its central function at the main victualling station for the Royal Navy, and its secondary role in the fitting out and laying up of warships.

At the beginning of the slaughtering season every autumn, the Commissioners would set up contracts stipulating that a certain number of pigs or cattle had to be slaughtered and delivered to each of the ports under their jurisdiction. Table 5.4 shows the quantities in question for Plymouth between 1756 and 1762. The table highlights an obvious development: from autumn and winter 1759–60, and for the following two years, no English beef was bought for Plymouth. In fact, a document dated 21 September 1759 explains that since English beef was considered to be excessively expensive, the London Commissioners opted for imports of Irish beef.[25] Consequently, the first

25 PRO, ADM 111/49, 21 September 1759.

contract with Irish suppliers was signed on 5 October for a delivery of 3,200 barrels of meat to Portsmouth and Plymouth.[26] In 1760 and 1761, 5,923 and 7,627 cwt, respectively, of Irish beef passed through Plymouth.[27]

Table 5.4 Cattle and pigs of English origin bought for the port of Plymouth

	Winter 1756–7	Winter 1757–8	Winter 1758–9	Winter 1759–60	Winter 1760–1	Winter 1761–2
Number of oxen	500	500	720	-	-	-
Number of pigs	600	1,000	-	> 500	-	1,000

Source: Table drawn up from data in the registers of the ADM 111 series.

A similar, although less marked trend occurred with pork. Irish pork was added to English supplies from April 1758[28] in the following quantities: 9,196 cwt in 1759; 8,260 cwt in 1760 and 7,891 cwt in 1761.[29] The third part of this book, dealing with markets, will list the geographical areas of Ireland that the cattle and pigs came from.

The acquisition of English pigs for Plymouth did not dwindle, however, despite the more competitive prices of the Irish suppliers. This was probably due to the fact that English pork was believed to preserve more easily and for longer, and therefore was often considered as being indispensable. In fact, any Irish pork on board warships setting off on long voyages was not supposed to exceed half of all the requirements in pork, and captains were instructed to have their men consume all Irish pork before eating pork of English origin.[30]

*

Supplies of malt, essential for the production of beer, were also quite frequently obtained by private contracts specifically drawn up for Plymouth. However, there was a difference between the malt market and the meat market (described above): in the latter, the contracts which were drawn up, despite being for a specific port, were nonetheless established under the jurisdiction of the London Commissioners. In contrast, the drawing up of contracts for malt came within the competence of the Plymouth Agent.

The more active role played by the yard's Agent in terms of contracts for malt was a direct consequence of the local nature of the market. Barley, from

26 PRO, ADM 111/49, 5 October 1759.

27 Statistics drawn up for the House of Commons: PRO, ADM 110/19, fos. 396–9.

28 PRO, ADM 111/49, 5 October 1759.

29 PRO, ADM 110/19, fos. 396–9.

30 PRO, ADM 111/48, 29 January 1759.

Table 5.5 The victualling of the port of Plymouth in malt in 1760 and 1761

Date	Supplier	Quantity in *quarters*	Price per *quarter* in *shillings/pence*
15 Feb. 1760	Mr Mooring	100 *qtrs*	22s/10d
7 Feb. 1760	Mr Thomas	400 *qtrs*	24s
7 March 1760	Mr Benness	?	24s
12 March 1760	Mr Benness	?	24s
31 March 1760	Mr Bunny	3,000 *qtrs*	24s
7 April 1760	Mr Blake	70 *qtrs*	24s
7 May 1760	Mr Mooring	1,000 *qtrs*	23s/5d
2 Jan. 1761	Mr Herbert	130 *qtrs*	24s
9 Jan. 1761	Mr Benness	150 *qtrs*	24s
4 May 1761	Mr Bunny	100 *qtrs*	24s
12 June 1761	Mr Benness	300 *qtrs*	23s/6d
24 June 1761	Mr Mooring	21 *qtrs*	23s/6d
28 August 1761	Mr Hammwick	100 *qtrs*	?
26 Oct. 1761	Mr Mooring	1,000 *qtrs*	23s/9d
20 Nov. 1761	Mr Tucker	500 *qtrs*	24s
20 Nov. 1761	Mr Bidwell	1,000 *qtrs*	25s
23 Nov. 1761	Mr Bidwell	500 or 1,000 *qtrs*	25s
2 Dec. 1761	Mr Whitehurch	100 *qtrs*	24s
11 Dec. 1761	Mr Hoskins	500 *qtrs*	24s
23 Dec. 1761	Mr Benness	400 or 500 *qtrs*	25s
23 Dec. 1761	Mr Cosmick	200 *qtrs*	24s
24 Dec. 1761	Mr Wood	450 *qtrs*	24s/5d

which malt is derived, was grown in great quantities in the south of England, and in particular in the environs of Plymouth. Documents reveal that many of Plymouth's suppliers of malt were local merchants who supplied the port directly: in 1760 and 1761 Messrs Mooring and Hoskins lived in Plymouth itself and Messrs Bunny and Whitechurch lived in Tavistock, while Mr Bennes was at 'Soton' (Southampton). It was more satisfactory for the Agent, rather than the Commissioners, to draw up the contracts. Nonetheless, the contracts could not be made binding until John Ommanney had received the Commissioners' approval concerning rates and quantities that had been negotiated by the Agent. In this respect, the Commissioners maintained total control over the market.

It is interesting to note that as a rule the prices for malt obtained directly by the Commissioners for the Victualling Board's requirements in London were higher than those negotiated by the Agent in Plymouth.

*

Until December 1759, all the Navy's requirements in wine passed through Guernsey. This was due to its fiscal status, for the island was the only destination at which wine, which was necessarily an imported product, could be unloaded without being taxed. Consequently, having been bought from foreign producers (see Part Three of this volume), wine was stocked in Guernsey and Agents in the various ports would arrange for it to be delivered directly from the island.

This complex system very frequently entailed long delays, for it either involved warships dropping anchor at Guernsey in order to load the wine, or the Agents arranging for victualling vessels to load the wine and then transfer it on board the squadrons, without the cargo ever touching dry land.

It was not long before the Commissioners began to complain about the futility of the system. The Admiralty intervened after their angry and loud protestations on 19 October 1759, because as of December 1759,[31] the Victualling Board was authorised to allow the ports of Plymouth and Portsmouth to be victualled directly in wine.[32] From then on and until the end of the war, Plymouth and Portsmouth received their supplies of wine directly from the merchants.[33] They were authorised to unload the wine in their naval yards without it being subject to tax, with the proviso that each Agent had to vouch that the pipes of wine be 'kept under the locks of the officers of the custom'. The Agent also had to ask for authorisation before every delivery.[34]

The London Commissioners who negotiated these special contracts had to first obtain the approval of the Lords of the Admiralty before they could sign the contracts.

Exceptional provisioning requirements

Last but not least in this study of provisioning strategies was the expedient adopted as a last resort, used only in times of great tension or pressing need. In this case, and following orders from London, the provisions were sent, with all due haste and in the quantities required, from other ports.

Michael Russell, from the port of Dover, was probably the most solicited Victualling Board Agent, and was regularly requested to supply not only Plymouth, but also Portsmouth in beer and bread. This was mainly due to the fact that Dover did not play a leading role in the victualling of troops during the Seven Years War, and consequently its production of victuals, which was seen as a precious reserve, was naturally sent to other ports which it could re-supply more quickly than could London. In June 1760, for example, Russell

31 PRO, ADM 110/19, 19 October 1759.

32 PRO, ADM 111/49, 3 December 1759.

33 PRO, ADM 111/49, 3 December 1759; ADM 111/50, 30 April 1760; ADM 111/51, 7 January and 8 April 1761; ADM 111/52, 18 November 1761; ADM 111/54, 4 August and 18 September 1762.

34 PRO, ADM 110/20, 21 April 1760.

loaded two brigantines, the *Robert* and the *John*, with victuals for Plymouth and Portsmouth respectively, having chartered them at the rate of 13 shillings per ton per month.[35] The Dover–Plymouth shipments were directly linked to those of Plymouth to the troops, so once again, speed and impeccable organisation were of the essence. On 22 May 1761, John Ommanney received orders to send cattle to Belle-Ile. Just three days later, Michael Russell in Dover had already been apprised by Plymouth of their request for 2,000 sacks of bread, as well as for quantities of beef and pork.[36]

Plymouth also occasionally received supplies from Chatham and, perhaps more surprisingly, from Portsmouth. Any rivalry that may have existed between the two yards in times of peace simply could not be allowed to manifest itself in times of war. Stocks of provisions had to be balanced, and there was no question of either port being in need of any sort. Consequently, 160 pipes of wine were sent from Portsmouth to Plymouth in March 1760: 'The Right Honourable Lords Commissioners of the Admiralty', wrote the Commissioners,

> approve to our sending either the *Liberty* and *Blacket* transport from Portsmouth with 160 pipes of wine to Plymouth to prevent any delays that may happen by unforeseen events to the ships ordered to fit out at that port and that it will be unnecessary to put any wine on board either the *Union* or *Neptune*.[37]

In August and October 1761, 100 quarters of malt were bought from Portsmouth and sent to Plymouth.[38] It was the port Agents' responsibility to inform the Commissioners of these transactions, so that a centralised accounts system could be maintained. In Plymouth, John Ommanney was required to send the Victualling Board Commissioners his weekly accounts so that a tight check could be kept on the port's stocks and reserves. It was absolutely essential to be able to calculate at any given moment the reserves of, for example, bread, butter or meat stocked in the warehouses.

Another reason for maintaining such tight control over the stocks was to avoid all wastage. In the event of an abundance of bread, the number of working ovens would be reduced. Sometimes, however, there was no plan to consume the reserves, and in this case the Agent would be asked to sell the victuals before they became inedible. The Plymouth Agent would occasionally organise public sales in which he would dispose of '*old stores, board and iron hoops*'. The money from these sales was destined for the Right Honourable

35 PRO, ADM 111/50, 13 and 16 June 1760.

36 PRO, ADM 111/51, 22 and 25 May 1761.

37 PRO, ADM 111/50, 26 March 1760.

38 PRO, ADM 111/54, 10 August and 26 October 1761.

George Grenville, Treasurer to the Navy until 1762, when Lord Barrington succeeded him.

These public sales did not occur with any defined regularity, but simply took place as often as necessary. They were an excellent means of clearing out the warehouses so that they could be re-stocked. John Ommanney organised between two and seven such sales every year between 1757 and 1762. There were usually at least two sales per year: one in the summer and one in the winter.

Table 5.6 A detailed example of a public sale organised by John Ommanney in Plymouth

	Quantity	Sales figures
Ox hides	77	£23/16/9¼
	314	£142/12/5
	37	£15/14/4½
	56	£29/19/2¼
Sheep skins	117	£5/17/0
	46	£1/14/6
	240	£6/0/0
	108	£1/7/0
Tallow		£8/0/11
		£100/16/9
		£205/10/9½
Bellies	54	£2/14/0
	78	£3/10/2¼
	8	£0/4/0
Offal		£15/10/11
		£557/8/9¾

Source: PRO, ADM 110/20, 24 August 1761.

It must be noted that spoiled flour was sometimes sold for consumption by prisoners of war. Ommanney earned £300 between 29 September and 31 October from the sale of such flour.

There was, consequently, very little waste. Any provisions that could not be used in victualling were sold, which helped a great deal in the management of expenses. In July 1761, for example, £1,369 had been made from sales in the different ports.

The next two chapters will focus on the secondary British ports and the outlying ports and stations. Since they had radically different management methods from the examples that have just been under scrutiny, it will be possible to appreciate the advantages and limitations inherent in the two systems. Plymouth's vocation to become one of the primary victualling yards was indeed confirmed during the war years. The base then went from strength to strength in the ensuing decades, culminating in the building of the imposing victualling complex at Stonehouse after the end of the Napoleonic Wars.

6

Victualling in the Secondary Home Ports and in the Overseas Stations

Unlike the primary yards which were placed under the direct jurisdiction of the Victualling Board, the Commissioners entrusted each secondary home or overseas port to a single contractor who had to be in a position to fulfil all and any demands for victuals made by the Royal Navy.

There were ten secondary home ports: Greenock, Whitehaven, Liverpool, Milford and Bristol on the west coast; Leith, Newcastle, Hull, Lynn and Yarmouth on the east coast. To these should be added five Irish ports – Londonderry (from September 1759), Belfast-Carrickfergus, Dublin, Waterford, Cork – and a dozen other overseas stations: in North America there was Louisburg (from July 1759), Quebec (from March 1761), Halifax in Nova Scotia, New York, and Charleston in South Carolina; in the West Indies, Barbados and the Leeward Islands, Jamaica and Havana (from October 1762); and finally Lisbon, Oporto, Madeira and Gorée.

The methods of victualling at these bases will be examined first, and this will be followed by an examination of the suppliers and their competitors, and a geographical breakdown of the markets supplying them. These indices will reveal not only the nature but also the importance of commercial networks in ensuring that the system worked optimally. Lastly, the chapter will focus on a specific trader who played a major role in Royal Navy victualling: the London merchant John Biggin.

How This Victualling Was Carried Out

The contractor of each station listed above had to be able to supply the vessels stationed there with a wide range of victuals and other goods. Whichever port was in question, the range of victuals was identical: the contractor had to be in a position to supply not only bread, peas, porridge, butter, cheese, vinegar and fresh beef, but also casks, biscuit bags and iron hoops for the barrels. The only variation occurred in terms of beverages: in the home, Irish and North American ports the contractor supplied beer, whereas ships putting in at Madeira, Lisbon and Gorée were supplied with wine and those putting in at the West Indian ports were supplied with rum.

The following procedures had to be followed in order for a contract to be drawn up: the Commissioners of the Victualling Board advertised in the press (usually in the *London Gazette*)[1] that on a specific date an invitation to tender bids for the victualling of a specific station would be made, and that merchants should make known their prices for each article or victual. If the invitation to tender bids concerned one of the Irish ports, then the Postmaster General at Dublin, the Collector of Customs at Cork and the Commissioner of the Revenue in Ireland were also notified:

> Let it be advertised and publication made as usual and send copys thereof and drafts of the contract to the Postmaster General at Dublin, and Collector of the Customs at Cork. And let the conditions of the contract be also sent to the secretary to the Commissioner of the Revenue in Ireland.[2]

There was a delay between the date when the decision to offer the contract was taken and the setting up of the contract itself, which could run from nine[3] to forty-two[4] days. The longest delays concerned the Irish ports, since they were dependent on the Commissioners opening the markets to dealers resident in Ireland. On the other hand, it is striking to note the much shorter delays for more distant yards: nine, twelve, twenty-eight, twenty-three and fourteen days respectively for those in Jamaica,[5] Madeira,[6] New York,[7] Nova Scotia[8] and Havana.[9] These shorter delays did not allow enough time for proposals to come through from the places in question and necessitated the presence in London, if not of the suppliers themselves, then at least a representative. This manner of proceeding illustrates the extent to which victualling was a highly concentrated operation. This aspect of victualling will be returned to in the second and third sections of this chapter.

Until February 1762 – that is to say for most of the duration of the war – the traders negotiated by letter. None of them appeared before the Victualling Board in person (see Table 6.1).

Nesbitt and Franks associates apparently put an end to the above practice by choosing to put in a personal appearance before the Commissioners on

1 PRO, ADM 111/44, 8 June 1756.

2 PRO, ADM 111/45, 22 November 1756.

3 PRO, ADM 111/44, 8 June 1756. Period of time relative to victualling in Jamaica.

4 PRO, ADM 111/45, 22 September 1757. Period of time relative to victualling in Waterford.

5 PRO, ADM 111/44, 8 June 1756.

6 PRO, ADM 111/48, 1 November 1758.

7 PRO, ADM 111/53, 15 January 1762.

8 PRO, ADM 111/53, 5 May 1762.

9 PRO, ADM 111/54, 8 October 1762.

26 February 1762. This was the date which had been set aside for choosing from the bids made for victualling the Royal Navy ships which were due to be stationed in Jamaica. This allowed the Commissioners, once they had heard all the bids, to ask the two associates whether they would agree to a certain discount. Nesbitt and Franks agreed and thus won the contract.[10] This most probably had the effect of encouraging other merchants to appear in person before the Victualling Board. Indeed, one of them appeared in person at the three last calls for bids of the war (see Table 6.1).

Generally speaking, the Commissioners did not choose their contractor on the day the bids were received.[11] The range of products was so vast that the Commissioners, at first sight, could not accurately assess the most competitive proposal. In order to do so, it was necessary to bring all the different bids to the same base, whilst taking into account the specific victualling needs for the geographic region in question.

The bids were then imparted to the Victualling Board's two accountants: Denham Skeet, accountant of cash, and Richard Henshaw, accountant for stores. Within the week they would work out, on the basis of a *per man a day* calculation, which merchant offered the most competitive terms.[12] This supplier usually won the contract, but from time to time, when two or more offers were substantially similar, the Commissioners would take into account other parameters such as whether the produce would be guaranteed by the merchant, and to what extent payment deadlines could be negotiated. On 1 August 1759, for example, John Biggin's rates *per man a day* were calculated as being 36/100th penny more than those of John Allen, and yet the Commissioners gave the contract to Mr Biggin thanks to the fact that he guaranteed the preservation of his supplies, and offered a six-month deferred payment proposal:

> The Board thinking the advantage the Crown must reap by having the provisions warranted and by six months credit instead of paying ready money would be more than the difference in the price, came to a resolution to agree with Mr. Biggin in case he would agree to take 3d/pound for fresh beef.[13]

Mr Biggin was nonetheless obliged to show goodwill by lowering his asking price for fresh beef on 8 August 1759.[14]

10 PRO, ADM 111/53, 26 February 1762.

11 Unless one of the merchants made a particularly competitive offer: PRO, ADM. 111/45, 2 March 1757.

12 Cf. the ADM 111 series.

13 PRO, ADM 111/49, 1 August 1759.

14 PRO, ADM 111/49, 8 August 1759.

Table 6.1 Bids pertaining to secondary home ports and outlying stations

Date	Victualling station	Number of merchants	Means of communication	Names	Chosen supplier
19 May 1756	Lisbon	5	By letter	Philip Jackson, George Amyand, Messrs Roberts and Baines, Anthony Bacon, Messrs Fowkes and Co.	Philip Jackson for fresh beef George Amyand for wine
19 May 1756	Yarmouth	4	By letter	Gerard Trotter, David Urquehart, Anthony Bacon, John Kerridge	David Urquehart
2 June 1756	Liverpool	1	By letter	Gildart for Lawrence Carr	Lawrence Carr
2 July 1756	Madeira	3	2 by letter 1 present (R. Baker)	Richard Baker (present), John Catanach, Richard Oswald	Richard Baker
14 July 1756	Bristol	1	By letter	Messrs Devonshire, Reeve and Lloyd	Messrs Devonshire, Reeve and Lloyd
11 August 1756	Barbados and the Leeward Islands	3	By letter	Samuel Fouchet, John Biggin, Auguste Boyd,	John Biggin
17 November 1756	Hull	3	By letter	John Biggin, John Kerridge, David Urquehart	John Kerridge
8 December 1756	Leith	5	By letter	William Alexander and Son, James Campbell, David Urquehart, James Riddell, John Biggin	John Biggin
2 March 1757	Belfast-Carrickfergus	2	By letter	Robert Gordon, John Biggin	John Biggin
22 April 1757	Milford	2	By letter	Jos. Allen, John Biggin	John Biggin
2 May 1757	Cork	2	By letter	Auguste Boyd, John Biggin	John Biggin
20 May 1757	Lynn	2	By letter	Georges Hogg, David Urquehart	David Urquehart
8 June 1757	Waterford	2	By letter	Auguste Boyd, Anthony Bacon	Auguste Boyd
13 June 1757	Whitehaven	1	By letter	Anthony Bacon	Anthony Bacon
19 August 1757	Dublin	3	By letter	Auguste Boyd, John Biggin, Messrs Clements and Evory	Messrs Clements and Evory
9 September 1757	Fresh beef for Lisbon	3	By letter	Georges Amyand, Philip Jackson, Anthony Bacon	Philip Jackson
23 December 1757	Nova Scotia	2	By letter	Peter Mauger, Samuel Fouchet	Peter Mauger
11 January 1758	Yarmouth II	2	By letter	David Urquehart, James Riddell	David Urquehart
29 March	Newcastle-Tinmouth Haven	2	By letter	John Biggin, James Riddell	James Riddell
21 April 1758	Hull II	5	By letter	John Biggin, James Hamilton, Thomas Humberstone, David Urquehart, Samuel Wallen	David Urquehart

28 April 1758	Barbados and the Leeward Islands II	2	By letter	John Biggin, Samuel Fouchet	Samuel Fouchet
9 October 1758	Milford II	4	By letter	John Allen, John Biggin, Messrs Garnar and William, David Urquehart	John Biggin
1 November 1758	Madeira II	4	By letter	Richard Baker, John Catanach, Matthew Clarmart, Robert Scott	Richard Baker
8 November 1758	Lisbon II	1	By letter	John Biggin	John Biggin
6 July 1759	Nova Scotia II	4	By letter	John Biggin, Sir Alexander Grant, Messrs Nesbitt and Franks, J. Hanbury	Sir Alexander Grant
13 July 1759	Louisburg	3	By letter	John Biggin, Sir Alexander Grant, Messrs Nesbitt and Franks	John Biggin
13 July 1759	Senegal-Gorée	1	By letter	John Biggin	John Biggin
7 March 1760	Barbados and the Leeward Islands III	1	By letter	John Biggin	John Biggin
9 April 1760	South Carolina	1	By letter	John Biggin	John Biggin
23 May 1760	Jamaica	4	By letter	John Biggin, Auguste Boyd, Samuel Fouchet, Messrs Nesbitt and Franks	Messrs Nesbitt and Franks
18 June 1760	Londonderry	2	By letter	John Biggin, James Evory and Co	James Evory and Co.
4 March 1761	Quebec	3	By letter	John Biggin, Robert Grant, Messrs Nesbitt and Franks	John Biggin
14 August 1761	Yarmouth III	3	By letter	James Riddell, David Urquehart, John Kerridge	David Urquehart
12 February 1762	Hull III	2	By letter	James Langton, David Urquehart	James Langton for fresh beef David Urquehart for sea provisions
12 February 1762	New York	3	By letter	John Biggin, Moses Franks, William Grant	William Grant
26 February 1762	Jamaica II	2	1 present 1 by letter	John Biggin (by letter), Messrs Nesbitt and Franks (present)	Messrs Nesbitt and Franks
15 March 1762	Madeira III	3	1 present 1 by letter	Richard Baker (present), John Biggin, John Pringle	John Biggin
2 June 1762	Nova Scotia III	4	1 present 3 by letter	John Biggin, William Grant (present), Joshua Mauger, John Mill	Joshua Mauger for fresh beef John Mill for sea provisions
22 October 1762	Havana	5	1 present 4 by letter	Anthony Bacon (present), Messrs Nesbitt and Franks, James Noble, Charles Ogilvie, James Riddell	Anthony Bacon for fresh beef Charles Ogilvie for sea provisions

If and when the going rate for a supplier's product was no longer competitive, the Commissioners would inform him of their decision to continue doing business with him provided he agreed to match his competitors' prices.[15]

Since the product in question was almost always fresh beef – a victual that was particularly expensive – the Commissioners always delayed making a definitive decision so that the competitors could play out the market amongst themselves. This eventually led to the separation of 'fresh beef' on a ship's victualling list from other victuals, commonly named *sea provisions*.

Even before the declaration of war, separate contracts had already been drawn up for the supply of fresh beef to Royal Navy vessels putting in at the two naval stations in the Caribbean: in Jamaica,[16] and in Barbados and the Leeward Islands.[17] On 12 February 1762, while the contract for the home port of Hull was being reviewed, the Commissioners decided to draw up two contracts: one for James Langton, supplier of fresh beef; the other for James Urquehart, supplier of all other provisions; for prices established with the two suppliers respectively.[18] However, the separate contracts drawn up in May and October 1762 for Nova Scotia[19] and Havana[20] were not established empirically, but were pre-arranged, and the suppliers were informed when the invitation to tender was announced.

Similarly, since wine was the main victual to be supplied to the King's ships putting in at Madeira, the Commissioners decided to step up the competition there and give themselves more leeway by announcing that they were giving themselves the possibility of: '*be[ing] at liberty to accept the offers of wine separately from the other provisions*'.[21]

Once a deal had been struck with a contractor, the latter had to write an affidavit in which he committed himself to respecting the clauses particular to the agreement in question and the regulations in force at that time.[22] Only then would a contract be drawn up. Also, the Victualling Board's stipulations aside, contractors whose victualling sources came from Ireland were also confronted with draconian legislation dating back to the first half of the eighteenth century and established by the Irish Parliament of that time with a view

15 PRO, ADM 111/47, 28 April 1758; ADM 111/49, 13 July and 8 August 1759; ADM 111/52, 19 August 1761.

16 PRO, ADM 111/44, 8 June 1756.

17 Notice the absence of fresh beef in the following contracts: 11 August 1756, 28 April 1758 and 7 March 1760.

18 PRO, ADM 111/53, 12 February 1762.

19 PRO, ADM 111/53, 5 May 1762.

20 PRO, ADM. 111/53, 8 October 1762.

21 PRO, ADM 111/48, 18 October 1758.

22 PRO, ADM 111/45, 19 November and 29 December 1756.

to improving the quality of meat and butter in order to encourage exports. The scholar John O'Donovan has reprinted these statutes which played a determining role in the positive development of Ireland's commercial balance.[23] These measures began by an Act in 1703 that forbade the slaughter of oxen before they had reached their fifth year, and which regulated the contents of casks. Amendments in 1723, 1725, 1739 and 1749 also obliged the suppliers to appoint two local men, known as *securitys*, who guaranteed that regulations would be strictly followed. The *securitys'* names were transmitted to the Commissioners of the Victualling Board.[24]

Regulations concerning the export of Irish butter were also subject to six successive Acts – namely those of 1679, 1707, 1715, 1717, 1721 and 1723. The 1723 Act, which applied until 1812, ruled that only oak, ash and sycamore wood could be used in making the barrels, and that each barrel had to be bound with a dozen hoops. The Act also stated that a public weightmaster should ensure that all barrels destined for export corresponded exactly to prescribed norms. The weightmaster received the sum of one penny per barrel for carrying out these duties.[25]

These contracts were drawn up for a minimum period of one year, at which point either party could end the agreement by giving six months' prior notice: '*12 months certain 6 months warning*'.[26] Eight months' notice was required for contracts dealing with Barbados,[27] and three months' for Quebec.[28] The only exception was for contracts for fresh meat destined for the victualling of ships in Jamaica. This was due to the fact that there were very low numbers of horned cattle in the West Indies and yet linked to this was a disproportionately greater demand for beef since large naval forces were stationed permanently in the West Indies.[29] In this specific context, contracts were drawn up for a minimum of four years, and one year's notice was required to terminate the contract.[30]

23 J. O'Donovan, *The Economic History of the Live Stock in Ireland* (Dublin and Cork, Cork University Press, 1940). See also D. Dickson, 'The Cork merchant community in the eighteenth century', in *Négoce et industrie en France et en Irlande*, ed. L.M. Cullen and P. Butel (Paris, CNRS, 1986); J.F. Shepherd and G.M. Walton, *Shipping, Maritime Trade and the Development of Colonial North America* (Cambridge, Cambridge University Press, 1972).

24 See for example PRO, ADM 111/45, 18 February 1757; ADM 111/50, 23 June 1760.

25 O'Donovan, *Economic History*, 128–9.

26 See the ADM 111 series.

27 PRO, ADM 111/47, 1 May 1758; ADM 111/50, 7 March 1760.

28 PRO, ADM 111/51, 4 March 1761.

29 Buchet, *La lutte*.

30 PRO, ADM 111/44, 8 June 1756; ADM 111/45, 18 February 1757; ADM 111/50, 4 June 1761.

None of these contracts was therefore for a fixed term – they continued until either the Victualling Board or the contractor saw fit to end them, always by respecting the prescribed period of notice. It must be underlined, however, that it was usually the Commissioners who ended a contract, often because another dealer would spontaneously approach them with highly advantageous terms. At this point, the Commissioners would publish an invitation to tender bids with a view to renegotiating the transaction. Reprinted below is one example from the Victualling Board registers:

> John Allen offer to contract for victualling His Majesty's ships at Milford on cheaper terms than the present contract. Let warning be given Mr. Biggin to determine his contract for victualling H.M.S. at Milford and acquainted Mr. Allen that we have given warning and shall publish to treat for a new contract of which notice will be given in the public papers.[31]

Sometimes it also happened that suppliers who were confronted by huge demand would slightly increase their prices, simply because they did not find it to their advantage to pursue their contract to its term. This was the case in Liverpool in 1756,[32] in Jamaica in 1762[33] and in Nova Scotia in 1762.[34]

In this way, certain ports or stations maintained the same contract throughout the duration of the war, whilst others, such as Yarmouth, Madeira and Nova Scotia, had up to three invitations to tender bids. Generally speaking, as Table 6.1 illustrates, the first year of the war witnessed the renegotiation of contracts for the home ports, whilst the succeeding years were more marked by the renegotiation of overseas contracts. This trend reflected the redeployment of British naval forces to overseas theatres, especially from 1758.

The Commissioners' task did not end with the granting of contracts to the most competitive suppliers. Similarly to the manner in which they proceeded with contracts granted in the ports directly under their jurisdiction (see Chapters 4 and 5), they were as careful concerning the secondary home ports and saw to it that the contractors honoured their commitments. They scrupulously calculated the sums deductible from their suppliers' bills for goods that had perished before the guaranteed date.[35] At the slightest sign of trouble, the supplier was summoned and was asked precisely how he planned to solve the problem.[36] On one particular occasion in December 1758, a supplier called

31 PRO, ADM 111/47, 21 July 1758; see also PRO, ADM 111/52, 31 August 1761.

32 PRO, ADM 111/45, 24 December 1756.

33 PRO, ADM 111/53, 6 January 1762.

34 PRO, ADM 111/53, 5 May 1762.

35 See for example PRO, ADM 111/52, 21 October 1761.

36 See for example PRO, ADM 111/48, 15 September 1758; ADM 111/52, 31 August 1761.

Peter Mauger, who was also the contractor for Nova Scotia, was taken to court for breach of contract.[37]

Sales and Marketing Networks

An exhaustive study of the ADM 111 archival series reveals the names of all the suppliers who succeeded in obtaining contracts, as well as those unfortunate competitors who did not. In an effort to better understand the underlying structure of this system and to see whether any networks existed, the ADM series provided enough information to carry out a breakdown of the declared capabilities of this group. The resulting information can be grouped into five geographical regions: Britain; Ireland; intermediary ports including Lisbon, Porto, Madeira, Senegal and Gorée; North America; and finally the West Indies.

Of the forty-three merchants who submitted tenders for one or more of the fifty-eight markets, only nine of them had the means to work beyond the immediate geographical vicinity. And of those, only two – Anthony Bacon and John Biggin – could operate on a truly worldwide basis. Two other merchants, Auguste Boyd and Samuel Fouchet, tried to obtain contracts in three geographical regions, and the remaining five could tender for contracts in two regions only.

The approach that has just been taken tends to hide the extremely concentrated nature of awards in this market. Reprinted in Table 6.3 are the names of those merchants who succeeded in winning a contract (or contracts) and their geographical region(s) of activity. The table reveals that twenty-one of the forty-three suppliers (48.84 per cent) shared the market between them. John Biggin alone obtained one-third of this market (18 contracts from a possible 58, i.e. 31 per cent). The five merchants who operated in several geographical regions monopolised more than half the market (31 contracts from a possible 58, i.e. 53.45 per cent). If to this sum is added David Urquehart's market share (the latter was the single most important supplier in Britain), then 62 per cent of all the contracts (36 out of 58) were shared between only six of the forty-three merchants who responded to the invitation to tender bids.

An examination of victualling in British ports reveals that there existed three categories of merchant: those who only operated in one port, those who operated in several ports but within one geographical region, and finally those who victualled the Royal Navy in two or more geographical regions.

Of the eighteen merchants who put in bids for the home ports (a low

[37] PRO, ADM 111/48, 11 December 1758.

Table 6.2 A breakdown of suppliers according to the number of regions in which they participated in the invitation to tender bids

Number of traders who participated in the invitation to tender bids in 5 geographical regions:			**2**
A. Bacon	J. Biggin		
Number of traders who participated in the invitation to tender bids in 4 geographical regions:			**0**
Number of traders who participated in the invitation to tender bids in 3 geographical regions:			**2**
A. Boyd (North America, West Indies, Ireland)			
S. Fouchet (North America, West Indies, Intermediary Bases)			
Number of traders who participated in the invitation to tender bids in 2 geographical regions:			**5**
J. Allen (Intermediary Bases, England)			
Messrs Evory and Clements (Intermediary Bases, Ireland)			
Messrs A., W. and R. Grant (North America, West Indies)			
Messrs Nesbitt and Franks (North America, West Indies)			
J. Riddell (West Indies, England)			
Number of traders who participated in the invitation to tender bids in 1 geographical region:			**34**
A. Baker	(North America)	J. Campbell	(England)
J. and P. Mauger	(")	L. Carr	(")
Hanbury	(")	Messrs Devonshire, Reeve and Lloyd	(")
J. Mill	(")	Messrs Garnar and William	(")
Ch. Olgivie	(West Indies)	J. Hamilton	(")
T. Fuller	(")	G. Hogg	(")
Attlays	(")	Humberstone	(")
J. Noble	(")	J . Kerridge	(")
G. Amyand	(Intermediary Bases)	J. Langton	(")
R. Baker	(")	G. Trotter	(")
J. Catanach	(")	D. Urquehart	(")
M. Clamart	(")	S. Wallen	(")
Fowkes	(")	G. Ward	(")
Ph. Jackson	(")	T. Wilson	(")
R. Oswald	(")	R. Gordon	(Ireland)
J. Pringle	(")	W. Kennedy	(")
Messrs Roberts and Baines	(")		
R. Scott	(")		
Total number of merchants:	**43**		

figure, reflecting the highly centralised nature of this victualling), thirteen of them belonged to the first group. They managed to obtain contracts in only two of the ten home ports (Liverpool and Bristol), except James Langton who only managed to obtain a contract for fresh beef in Hull (in February 1762 – see Table 6.3).

In Liverpool, a local merchant called Laurence Carr was alone in placing a bid in June 1756. A certain Mr Gildart[38] transmitted his letter to London. He was succeeded by Thomas Wilson, another local merchant.[39] The situation in

38 PRO, ADM 111/44, 2 June 1756.

39 PRO, ADM 111/54, 19 November 1762.

Table 6.3 Merchants in the ten home ports

Merchants who participated in the invitation to tender bids at only one home port: 13

Merchants	*Ports*	*Contracts won*
Messrs Devonshire, Reeve and Lloyd	Bristol	1
James Hamilton	Hull	0
Humberstone	"	0
James Langton (fresh beef only)	"	1
Samuel Wallen	"	0
George Ward	"	0
James Campbell	Leith	0
Lawrence Carr	Liverpool	1
Thomas Wilson	"	1
George Hogg	Lynn	0
John Allen	Milford	0
Messrs Garnar and William	"	0
George Trotter	Yarmouth	0

Merchants who participated in the invitation to tender bids at more than one home port: 3

Merchants	*Ports and participation in the invitations to tender bids*			*Number of contracts obtained*	
John Kerridge	Hull	1	**4**	1	**2**
	Yarmouth	3		1	
James Riddell	Hull	1	**4**	0	**1**
	Leith	1		0	
	Newcastle	1		0	
	Yarmouth	1		1	
David Urquehart	Hull	3	**8**	2	**5**
	Leith	1		0	
	Lynn	1		1	
	Milford	1		0	
	Yarmouth	2		2	

Merchants who participated in parallel invitations to tender bids in one or more other geographical regions: 2

Merchants	*Ports and participation in the invitations to tender bids*			*Number of contracts obtained*	
Anthony Bacon	Whitehaven	1	**2**	1	**1**
	Yarmouth	1		0	
John Biggin	Greenock	1	**10**	1	**7**
	Hull	3		1	
	Leith	2		2	
	Milford	2		2	
	Newcastle	2		1	

Total number of contracts: 20

Total number of merchants: 18

Bristol did not show any signs of greater competitiveness either: the market was awarded in July 1756 to three merchants who formed an association just for the occasion and who easily won the contract – granted in the absence of any other proposals.[40]

There were three merchants in the second category: Messrs John Kerridge, James Riddell and David Urquehart. The latter, who was the most powerful, bid on eight separate occasions and obtained five contracts: two for Yarmouth, two for Hull and one for Lynn. These ports were all fairly close to one another and reflect the extent to which Urquehart preferred to operate within a limited area – in this case, Norfolk. A letter dated 31 January 1757 seems to indicate that he was from Yarmouth.[41] In this perspective it is noticeable that only one of his eight bids concerned a west coast port, and that was for Milford.

It seems that John Kerridge was one of his direct competitors, for he engaged in the battle for Yarmouth and Hull on a number of occasions. These were contracts that he had obtained, in fact, at the outbreak of hostilities. On occasion, competition between the two men could become very harsh. In May 1756, when Kerridge had just lost the Yarmouth contract to Urquehart, the former revealed his bitterness to the Commissioners by reminding them that he had always looked after the Crown's best interests.[42] The Commissioners condemned him out of hand, replying that: '*it being the standing rule of this Office on making contract to agree with the lowest tenderers*'.[43] In August 1761, during another invitation to tender bids for Yarmouth, John Kerridge put forward a proposal even though earlier his company had been declared bankrupt (in 1752). David Urquehart did not lose a minute in writing to the Commissioners and informing them that since John Kerridge did not possess a certificate from the court of Bankruptcy, he could not participate in the bid for tenders.[44] This communication put the Commissioners in a very awkward position, and they consequently sought legal advice.[45] Since this came down in favour of Kerridge,[46] Urquehart had to lower the price of his fresh beef in order to be sure to retain his contract.[47]

40 PRO, ADM 111/44, 14 July 1756. See also ADM 111/44, 8 June 1756.

41 PRO, ADM 111/45, 31 January 1757.

42 PRO, ADM 111/44, 27 May 1756.

43 PRO, ADM 111/44, 27 May 1756.

44 PRO, ADM 111/52, 19 August 1761.

45 PRO, ADM 111/52, 19 August 1761.

46 'The Board may legally contract with a person that has become bankrupt although he should not have obtained his certificate, but apprehends it would be advisable in such case to take security.' PRO, ADM 111/52, 21 August 1761.

47 PRO, ADM 111/52, 21 August 1761.

A few days later, an embittered John Kerridge informed the Commissioners of his intention to offer to victual, at a lesser price, any vessels which requested this service, in Hull.[48] The Commissioners responded by breaking their contract in this port with David Urquehart and once again issuing an invitation to tender bids.[49]

James Riddell submitted four bids (one in Leith, one in Newcastle, one in Hull and one in Yarmouth). Of these, only the bid for Newcastle, in which he was in competition with John Biggin, was successful.[50] Since there is a lack of archival material, one can simply notice that this merchant placed all his bids for ports on the east coast, thus possibly confirming his desire to concentrate his efforts on this coast. What is known is that he had an agent, Armorer Donkin, who was in the habit of going to North Shields to buy the provisions and barrels necessary to victual the Royal Navy's ships in Newcastle.[51]

The remaining two contractors operated on a much larger scale and their work included Irish and overseas ports. The first of these two, Anthony Bacon, had attempted to obtain a contract in Yarmouth and had failed due to the intensity of the local competition (see above).[52] However, in June 1757 he obtained, without any difficulty whatsoever, the contract for Whitehaven. In fact, he had been the only merchant to place a bid,[53] and furthermore was very well established in Belfast where he was one of the biggest meat merchants. Significantly, Belfast was directly opposite Whitehaven.

Last but not least, John Biggin was, without a doubt, the most important supplier. He succeeded in obtaining no less than seven of the twenty available home markets and reigned supreme throughout the duration of the hostilities, victualling Greenock and Milford on the west coast, and Leith on the east coast; and from time to time he obtained contracts for Newcastle and Hull. A measure of this man's importance is that he failed to obtain a contract on only three out of a possible ten occasions, namely in Hull in November 1756 and April 1758 and in Newcastle in March 1758, and each time against very well established merchants – James Riddell and David Urquehart. The third part of this chapter will focus closely on Mr Biggin.

The victualling of Royal Navy ships anchoring in the five Irish ports, each of which was under the control of a contractor, involved an even more concen-

48 PRO, ADM 111/52, 31 August 1761.

49 The contract was once again given to David Urquehart, on 22 February 1762. At this point, John Kerridge did not place a bid. It seems, therefore, that Urquehart finally succeeded in ousting his rival.

50 PRO, ADM 111/47, 29 March 1758.

51 PRO, ADM 111/54, 1 October 1762.

52 PRO, ADM 111/44, 21 May 1756.

53 PRO, ADM 111/46, 13 June 1757.

trated set of circumstances (see Table 6.4). The Irish contracts were shared between only six merchants, and in fact three of these, namely John Biggin, James Evory and Auguste Boyd, dominated the others.

Table 6.4 Suppliers to the five Irish ports

Merchants	*Ports and participation in the invitation(s) to tender bids*			*Contracts obtained*	
Anthony Bacon	Waterford		1		0
John Biggin	Belfast-Carrickfergus	1	4	1	2
	Cork	1		1	
	Dublin	1		0	
	Londonderry	1		0	
Auguste Boyd	Cork	1	3	0	1
	Dublin	1		0	
	Waterford	1		1	
James Evory	Dublin	1	2	1	2
	Londonderry	1		1	
Robert Gordon	Belfast-Carrickfergus		1		0
William Kennedy	Londonderry		1		1
	Contracts awarded 6				
	Total number of merchants 6				

John Biggin, the most powerful of these men, placed bids for four of these five Irish ports and succeeded in obtaining the two most lucrative markets, namely Cork and Belfast-Carrickfergus. James Evory, who also happened to be the main supplier of meat to all the home ports under the direct jurisdiction of the Victualling Board, obtained Dublin and Londonderry. Auguste Boyd only placed bids for those ports situated on the southern half of the east coast of Ireland (Dublin, Waterford and Cork), and obtained a contract for the victualling of Royal Navy ships in Waterford. This port was centrally placed in relation to Dublin and Cork and was the only port that John Biggin did not bid for.

The three remaining suppliers played an insignificant role. Anthony Bacon lost Waterford to Auguste Boyd; Robert Gordon (who would not be heard from again) lost Belfast-Carrickfergus to John Biggin; and William Kennedy obtained a contract for Londonderry but only kept it for a year.

The above list reflects the near-impossibility for small, local merchants to establish themselves. Indeed, it is only because Londonderry did not benefit from the presence of a Victualling Board contractor until September 1759 that William Kennedy, a small-scale local merchant, managed to obtain his contract. Captain Samuel Hood (later Admiral Lord Hood), who had re-stocked his vessel when in Londonderry with beer and other provisions supplied at very modest prices by Mr Kennedy, had in fact recommended him to the Commissioners, who then decided, on 12 September 1759, quite simply to offer him a contract without issuing a general invitation to tender bids.[54]

However, once John Biggin and James Evory had been informed of this transaction, they immediately responded by proclaiming that they would victual the port at a lesser cost. Consequently, on 8 June 1760, James Evory, who already victualled vessels putting in at Dublin (a port which, because of its low prices, was particularly advantageous to the contractor[55]), won the bid for Londonderry. In fact, William Kennedy did not even bother to participate in the invitation to tender bids.[56]

To conclude, the victualling of the Irish ports contracted by the Victualling Board was placed entirely in the hands of important merchants whose intention was to partake in commerce on a worldwide basis. For these men (John Biggin, James Evory, Auguste Boyd *esq.*[57] and to some extent Anthony Bacon), Ireland, with its very large herds of cattle, was an essential starting point for obtaining future contracts.

The victualling of the North American naval stations (Nova Scotia (Halifax), Louisburg, New York, Quebec and South Carolina (Charleston)) proceeded in a similar fashion to that of the Irish ports. A mere four merchant companies shared the entire market, albeit very unequally (see Table 6.5).

John Biggin was the sole merchant to offer his services at all five stations and won three of the nine markets contracted out during the war (Quebec, Louisburg and South Carolina). A family company run by Sir Alexander Grant[58] and his kinsmen Robert and William failed to obtain Quebec and Louisburg but did succeed in gaining the markets of Nova Scotia and New York in 1759 and 1762 respectively. To this accomplishment can be added the partnership between the Grants and John Mill[59] who together succeeded in obtaining the market for sea provisions for Nova Scotia when it became available again in 1762.

Peter and Joshua Mauger only competed for the Halifax, Nova Scotia, market, and succeeded in obtaining two contracts, one for the victualling of

54 PRO, ADM 111/49, 12 September 1759.

55 PRO, ADM 111/49, 12 September 1759.

56 PRO, ADM 111/50, 18 June 1760.

57 PRO, ADM 111/50, 15 February 1760.

58 A self-made man, Sir Alexander came into his baronetcy in 1755 and, having made his fortune from Jamaican plantations, returned to London as a merchant with trading interests in the Mediterranean, West Indies, America, Africa (as a slaver) and India. He became MP for Inverness Burghs 1761–8 (PRO, ADM 111/51, 24 June 1761; D. Hancock, 'Grant, Sir Alexander, fifth baronet (1705–1772)', *Oxford Dictionary of National Biography* (Oxford, Oxford University Press, 2004).

59 John Mill brought over most of his provisions from Ireland (PRO, ADM 111/54, 20 August 1762).

Table 6.5 Suppliers to the five North American yards

Merchants	*Yards and participation in the invitation(s) to tender bids*			*Contracts obtained*	
Anthony Bacon	New York		**1**		**1**
John Biggin	Louisburg	1	**5**	1	**3**
	New York	1		0	
	Nova Scotia	1		0	
	Quebec	1		1	
	South Carolina	1		1	
Samuel Fouchet	Novia Scotia		**1**		**0**
Sir Alexander, Robert and William Grant	Louisburg	1	**5**	0	**3**
	New York	1		1	
	Nova Scotia	2		2	
	Quebec	1		0	
J. Hanbury	Nova Scotia		**1**		**0**
Joshua and Peter Mauger	Nova Scotia		**2**		**2**
Messrs Nesbitt and Franks	Louisburg	1	**4**	0	**0**
	New York	1		0	
	Nova Scotia	1		0	
	Quebec	1		0	
	Contracts offered 9				
	Total number of merchants 7				

the port in its entirety in December 1757, and a second one for fresh beef in June 1762. Anthony Bacon – the exporter of Irish meat – obtained the victualling of vessels putting in at New York for most of the war's duration.

Three other suppliers or companies participated in the invitation to tender bids but without the least success. These were J. Hanbury, who was one of the most important ship owners of the period, Samuel Fouchet, and finally the Nesbitt and Franks Company, which tendered bids for every market except South Carolina.

In addition to the fact that those involved in victualling these markets numbered a select few, the example of relatively important merchants such as the Grants, Maugers and Nesbitt and Franks, whose names did not appear in relation to victualling in the home areas and Ireland, illustrates the extent to which they operated in highly specific geographical areas. A review of victualling in the West Indies will throw further light on this tendency.

The West Indies practised extensive sugar cane farming and were dependent on external markets for their victualling needs. The ways in which this commerce operated have been carefully documented and researched by James F. Shepherd and Gary M. Walton and by Richard B. Sheridan in his study of the economy of the British West Indies.[60] The fact that the West

60 J.F. Shepherd and G.M. Walton, *Shipping, Maritime Trade and the Economic Development of Colonial North America* (Cambridge, Cambridge University Press, 1972); R.B. Sheridan,

Indies were positioned between North America and Europe, and that the prevailing winds and currents rendered a more direct route impassable to the Navy's sailing ships, meant that the West Indies became an active centre of Atlantic trade.[61] As a consequence, the profile of merchants offering to victual the numerous war ships at anchor in the West Indies was particularly cosmopolitan. There were two stations: Barbados and the Leeward Islands; and Jamaica. By the end of the war, the base at Havana had been added to the list (see Table 6.6).

Table 6.6 A list of merchants victualling the West Indies (sea provisions)

Merchants	*Stations and participation in the invitation(s) to tender bids*			*Contracts obtained*	
Anthony Bacon	Havana		**1**		**0**
John Biggin	Barbados and the Leeward Islands	3	**5**	2	**4**
	Jamaica	2		2	
Auguste Boyd *esq.*	Jamaica		**1**		**1**
Samuel Fouchet *esq.*	Barbados and the Leeward Islands	3	**4**	2	**2**
	Jamaica	1		0	
William Grant	Jamaica		**1**		**0**
Messrs Nesbitt and Franks	Jamaica	2	**3**	2	**2**
	Havana	1		0	
Charles Ogilvie	Havana		**1**		**1**
James Riddell	Havana		**1**		**0**
	Contracts awarded 10				
	Total number of merchants 8				

Of the eight merchants participating in the invitation to tender bids for sea provisions, seven of them held at least one contract in a different geographical area – the only exception being Charles Ogilvie.

An examination of the two main markets, namely Jamaica and Barbados and the Leeward Islands, reveals that all five merchants (John Biggin, Auguste Boyd, Samuel Fouchet, Nesbitt and Franks and William Grant) who were in competition for victualling contracts were already well established in North America. Three of these tradesmen also did business in Ireland: Biggin, Boyd and Fouchet.[62] Fouchet did not actually participate in the invitation to tender bids for the victualling of Irish ports but did put forward a proposal, in 1758,

Sugar and Slavery: An Economic History of the British West Indies 1623–1775 (Barbados, Caribbean University Press, 1974).

61 D.A. Farnie, 'The commercial empire of the Atlantic, 1607–1783', *Economic History Review*, 2nd series, XV (1962–3), 205–18.

62 PRO ADM 111/44, 11 August 1756.

to supply Plymouth, Portsmouth and Spithead with Irish pork.[63] Furthermore, in 1756, he supplied Gibraltar with its requirements in beef and pork originating from Ireland.[64]

As far as market share is concerned, Biggin and Fouchet both obtained two of the four contracts concerning Barbados and the Leeward Islands. Nesbitt and Franks, who passed their supplies via New York, took Jamaica from Auguste Boyd by acquiring two consecutive contracts (in 1760 and 1762).[65]

William Grant and Nesbitt and Franks, who had business connections in North America but not in Ireland, only offered their services for Jamaica, which was geographically much closer to the thirteen colonies. As Professor Pierre Chaunu[66] has pointed out, Barbados was surprisingly close to Europe since the trade winds allowed sailing ships to travel up to 398 km (247 miles) in twenty-four hours.[67] If, when taking into account the time/distance ratio, the Barbados-Jamaica passage was extremely rapid, it must be remembered that the same trip undertaken in the opposite direction in an era of sail-power was long and tricky. It is therefore logical that merchants whose supplies came from North America should limit their sphere of activity to Jamaica, and that Barbados and the Leeward Islands should be victualled more easily by vessels coming from Europe. These ships would deliver their cargo and then load up with products from the West Indies, transit to North America where they could again do commerce and then turn back towards the 'Old Continent'.

Since Jamaica had indigenous herds of cattle, the fresh beef markets were awarded to local merchants: George Barclay and Stephen Attlays. Barbados, however, had no specific contract. Havana did have a specific contract for the victualling of fresh beef, which was awarded to Anthony Bacon in 1762.

This fifth and last geographical area has been given the generic term of 'intermediary bases' due to the ports' staging post position on the routes to the Americas and the Indies. The Commissioners in London appointed contractors for the victualling of stations at Lisbon, Madeira, Oporto, and Senegal and Gorée (see Table 6.7).

Once more, John Biggin can be seen to have dominated the commercial scene: apart from the instance of wine in Madeira, he possessed the totality of the other four markets. If he only obtained the Madeira market in March 1762, it is undoubtedly due to the fact that before this date wine was not subject to separate contracts. In these circumstances, Richard Baker, one of

63 PRO ADM 111/47, 14 April 1758.

64 PRO ADM 111/45, 29 December 1756; 18 February and 2 March 1757.

65 PRO, ADM 111/53, 22 March 1762.

66 H. and P. Chaunu, *Séville et l'Atlantique 1504–1650* (Paris, SEVPEN, 1955–60).

67 Buchet, 'Des routes maritimes Europe-Antilles'.

Table 6.7 List of merchants tendering sea provisions for the four intermediary bases

Merchants	*Yards and participation in invitation(s) to tender bids*			*Contracts obtained*	
John Allen	Oporto	1		0	
George Aymand	Lisbon	1		1 (for wine only)	
Anthony Bacon	Lisbon	1		1 (salt meat only)	
Richard Baker *esq.*	Madeira	3		2 (victualling total) 1 (for wine only)	
John Biggin	Lisbon	1	4	1	4
	Madeira (excepting wine)	1		1	
	Oporto	1		1	
	Senegal and Gorée	1		1	
John Catanach	Madeira	2		0	
Matthew Clarmart	Madeira	1		0	
Messrs Fowkes and Co.	Lisbon	1		0	
Richard Oswald	Madeira	1		0	
John Pringle	Madeira	1		0	
Messrs Roberts and Baine	Lisbon	1		0	
Robert Scott	Madeira	1		0	
Contracts awarded 9					
Total number of merchants 12					

the Navy's main suppliers of wine, easily obtained the first two contracts, especially since wine was the main article provided in the island.[68]

The third and final part of this chapter will be devoted to a study of this most exceptional contractor, John Biggin.

John Biggin: Merchant on a Worldwide Scale

In an effort to add to any existing information on John Biggin – a merchant who obtained almost a third of all contracts for the victualling of vessels in the secondary home ports and the overseas ports and stations – it was decided to direct research efforts into an examination of archival material focusing directly on the man and his business network.

Unfortunately, no personal files on Mr Biggin have been conserved in the Public Record Office or in the National Maritime Museum. The next attempt

68 Buchet, 'Des routes maritimes Europe-Antilles'.

involved consultation of the London archives conserved by the London Chamber of Commerce. This route was chosen because it is known that John Biggin lived in London or its suburbs and often appeared before the Commissioners of the Victualling Board. He would also have had to enrol in the registers of the Corporation of London Merchants in order to be allowed to trade in the town. The heavy volumes of the *City Freedom Admissions* series list, in alphabetical order, the names of all these merchants together with the date on which they received their certificates. These certificates include some information of a more personal nature, such as the person's place of birth and father's profession.

The merchants' names only appear once in the *City Freedom* registers,[69] namely on the date on which they entered the corporation. In the case of John Biggin, it was necessary to start in 1756 (the year in which it seems he obtained his first contract), and then to work backwards. However, no trace of this supplier could be found, even as far back as 1701. It thus appeared that he had never joined the Corporation of London Merchants. Similarly, the Chamber of Commerce Library in London also had no mention of his name. The trail then moved to the National Archives in Dublin, which proved to be equally unfruitful.[70]

All that remained at this point was to refer back, once again, to the fragmentary and scattered information to be found in the Navy (ADM) series of the Public Record Office. This would at least present a picture of his various activities, and their geographical breakdown, and by comparing prices on different contracts, it would be possible to reconstruct his business network.

Additional activities

An important point concerning John Biggin's career that has not yet been referred to is that the archival series ADM 110 and 111 reveal that Biggin was probably one of the most important ship owners and ship brokers of the period.[71] As well as running his own transportation business, it appears that on a number of occasions he came to the Victualling Board's aid by providing them with freight transportation. He and his competitor John Hutchinson were the two ship owners most capable of assembling the greatest tonnage. For example, it was Biggin who saw to the Victualling Board's freight needs for the victualling of Gibraltar in 1756,[72] 1760[73] and 1761.[74] At the same time

69 Except in cases of being struck off, or other exceptional circumstances.

70 The author wishes to thank Aideen M. Ireland, archivist.

71 Syrett, *Shipping and Military Power in the Seven Years War*, 40–1.

72 PRO ADM 111/44, 20 August 1756.

73 PRO ADM 111/50, 22 August 1760.

74 PRO ADM 111/52, 22 September 1761.

in 1761, sixteen other ships of his fleet[75] were ferrying back and forth between Cork and the French coast in order to victual the troops stationed on Belle-Ile with livestock and green vegetables. On top of this, he still found the means to offer the services of six other vessels, totalling 1,420 tons, for foreign service.[76]

Printed below is a list of the names and tonnage of some of John Biggin's ships:

Achille: 340; *John and Dorothy*: 269; *Ann Galley*: ?; *Judith*: 369; *Antelope*: 287; *Lady Ann*: 250; *Arnold*: 400; *Lyde*: 356; *Centurion*: ?; *Mary*: 310; *Diamond*: 327; *Mary Ann*: 200; *Dorothy*: 190; *May Flower*: ?; *Nelly*: 222; *Pitt Galley*: 220; *Royal Duke*: ?; *Royal George*: 623; *Ruby*: 331; *St James*: 350; *Sea Venture*: 291; *Southampton*: ?; *Thomas and Richard*: 380.

The fact that Biggin was capable of readily and rapidly supplying vessels of great tonnage capacity placed him in a very favourable position with the Victualling Board. Consequently, when in June 1758 the Victualling Board had to urgently supply Commodore Howe's squadron stationed just off the French coast with provisions and miscellaneous products representing a volume of over 2,000 tons, the Commissioners did not hesitate to contact John Biggin directly.[77]

In the same way, Biggin was approached and asked to ensure the victualling of the fleet at Quiberon with livestock and fodder, potatoes and onions: 'the oxen must be fattest and the potatoes, onions and fodder the best that can be procured'.[78] This marked the beginning of an arrangement between Biggin and the Victualling Board in which he ensured the transport of livestock and fresh vegetables from Cork and Kinsale to the western squadron and to troops disembarking on Belle-Ile.[79] This enterprise began in 1761 and continued uninterrupted until the end of hostilities, and mobilised at least sixteen transport vessels.[80]

Biggin was reimbursed for the invoices he presented relating to the products he bought in Ireland, and then, based on forms that had to be signed by the Royal Navy captains upon receipt of the provisions, he also received a commission of £2 10s for every hundredweight of cargo carried in his ships.[81]

In an attempt to limit the financial disbursements that the Crown would have to make, Biggin was instructed, on every trip, to bring back the skins and

75 PRO ADM 111/52, 11 September 1761.

76 PRO ADM 111/52, 28 October 1761 and PRO ADM 111/53, 25 August 1762.

77 PRO ADM 111/47, 28 June 1758 and ADM 111/53, 26 May 1762.

78 PRO ADM 111/50, 22 February, 18 June and 18 July 1760.

79 PRO ADM 111/52, 27 July, 5 and 26 August 1761; ADM 111/54, 10 September 1762.

80 PRO ADM 111/52, 4 and 11 September 1761; ADM 111/53, 6 January 1762.

81 PRO ADM 111/50, 22 February 1760; ADM 111/54, 7 and 21 July 1762.

fat resulting from the slaughter of animals since his last voyage, to sell them at the current market price and to deduct this sum from the amount that was owed to him by the Victualling Board.[82]

In December 1755 and April 1760, Biggin tendered a bid to provide Portsmouth and Plymouth with wine, a gesture that was directly related to the fact that he owned a fleet of ships. He assured the Commissioners that he was equipped to deliver any quantity whatsoever of red or white Spanish wine.[83] As it happens, this contract was not awarded to him – it went to importers of Malaga, Vidonia and Fayol wines – but he would soon become one of the Navy's most important suppliers of brandy.

On 15 December 1758, he acted as an intermediary for a friend residing in Guernsey who had a French prize carrying brandy which the Victualling Board was willing to buy at the price of 3s 3d per gallon.[84] At the new year period 1760/1, Biggin proposed on his own account to provide 35,000 gallons of brandy at Guernsey, again at the rate of 3s 3d per gallon.[85] The Board accepted. Two months later, he obtained two further contracts for 40,000 gallons each: one was for a delivery of French brandy for Guernsey at a cost of 2s 8d per gallon, the other for Flemish brandy for Portsmouth at the rate of 2s 7d per gallon.[86]

The following year he proposed to deliver to Guernsey up to 80,000 gallons of brandy at the rate of 3s ½d a gallon.[87] The deal was accepted[88] and on 24 February 1762 Biggin sent the Victualling Board a certificate attesting to the delivery at Guernsey of 63,593 gallons of French brandy.[89] Consequently, due to contacts Biggin obviously had with French merchants even during the war years, probably thanks to his agent in Guernsey, Mr Le Marchant,[90] Biggin became the Navy's main supplier of brandy. Furthermore, the demand for brandy grew and grew as a direct consequence of naval operations in overseas theatres since beer did not keep for more than a few weeks at sea.[91]

Finally, it is worth noting that on 28 September 1759, John Biggin participated in bids made for the acquisition of a contract for the victualling of

82 PRO ADM 111/50, 22 February 1760; ADM 111/54, 7 and 21 July 1762.

83 PRO ADM 111/49, 3, 17 and 31 December 1759; ADM 111/50, 30 April 1760.

84 PRO, ADM 111/48, 15 December 1758 and 15 January 1759.

85 PRO, ADM 111/51, 2 January 1761.

86 PRO, ADM 111/51, 27 and 28 February, 3, 16 and 25 March 1761.

87 PRO, ADM 111/53, 11 January 1762.

88 PRO, ADM 111/53, 13 January 1762.

89 PRO, ADM 111/53, 24 January 1762.

90 PRO, ADM 111/52, 7 September 1761.

91 Buchet, 'Des routes maritimes Europe-Antilles'.

the ports of Plymouth and Portsmouth with fresh meat. However, since the Commissioners considered that all the prices tendered were too high, the market was not allotted to anyone.[92] On 20 February 1761, Biggin took it upon himself (there was no invitation to tender bids at the time) to propose delivery to England of 700 *barrells* and 120 *tierces* of Irish beef, as well as 500 *barrells* of pork. However it would appear that this offer was not taken up either.[93]

On a more personal level, it would seem from Biggin's frequent appearances before the Commissioners, and from the fact that letters which he sent to the Victualling Board were entered in their registers on the same day that they were written,[94] that he probably lived in Deptford or its immediate surroundings. Consequently, he must have managed his business affairs from London. This implies that in each yard with which he held a victualling contract, and which at the same time furnished him with other supplies, he had to rely on the agent that he had placed there.[95] A close examination of Biggin's correspondence reveals the names of a certain number of these agents and where they operated – permitting a greater understanding of the regional scope of his business. The following is a list of these agents: Charles O'Hara in Dublin; Scott in Leith; Ardouin, William Coates and John Tighe in Cork; Le Marchant in Guernsey; Henry White in New York; Francis Sench in Philadelphia; and Meyler, Halls and Torry in Barbados. These employees had to be extremely versatile and regularly send Biggin an account of their activities and any problems they may have encountered.

Even though a study of archival and other material has revealed hundreds of recordings of Biggin's name, on only one occasion was he given the title of esquire.[96] However, his social standing from an economic point of view was sufficiently elevated that on a number of occasions he directly addressed the Lords of the Admiralty himself.[97]

An attempt at reconstructing Biggin's business network

Since there remains, unfortunately, no trace of the correspondence exchanged between Biggin and his agents – indices which would no doubt have provided a means of reconstructing exactly how his business network functioned – a

92 PRO, ADM 111/49, 28 September 1759.

93 PRO, ADM 111/51, 20 February 1761.

94 PRO, ADM 111/47, 28 June 1758; ADM 111/50, 26 March, 30 June and 28 August 1760; ADM 111/51, 25 March 1761; ADM 111/53, 6 January 1762.

95 PRO, ADM 111/44, 9 September 1756; ADM 111/47, 10 May 1758; ADM 111/50, 5 September 1760; ADM 111/51, 25 March 1761; ADM 111/54, 3 December 1762.

96 PRO, ADM 111/52, 17 August 1761.

97 PRO, ADM 111/53, 21 April 1762.

correlation of his contracts has been undertaken in order to reveal his operating methods.

Table 6.8, which offers a sequential presentation of the various contracts obtained by John Biggin, also reveals his main bases and sources of supply. It becomes clear that throughout the entire war period Biggin always victualled at least one port in three of the five geographical regions (namely, Britain, Ireland and North America). Biggin reigned supreme in Leith and Greenock (the former on Scotland's east coast, the latter on its west coast); in Cork and Belfast-Carrickfergus, where he obtained his contract on 2 March 1757 and never lost it again; and in South Carolina in America. It is not clear whether he kept these contracts because the Victualling Board was not inclined to send out new invitations to tender bids, or whether he succeeded in several successive adjudications. It would appear to be quite plausible that in those ports in which he held well-established contracts he was also extremely well known to local producers. These connections would have been the pivotal point from which Biggin could have easily and cheaply set up an export business, thus enabling him to win other contracts.

By drawing on British and Irish resources, he obtained contracts in three of the four intermediary ports, and by March 1762 in all four of them, plus a few in the West Indies, to which vessels were rapidly dispatched thanks to the trade winds. Documents have revealed that the beef, pork and butter that Biggin was delivering to Lisbon, Oporto, Barbados and the Leeward Islands came from Ireland,[98] and that the barley, peas and porridge oats originated from Britain.

Biggin's ships would then continue on their journey until they reached South Carolina, which was Biggin's main centre of activity in North America. South Carolina was geographically extremely well-positioned, and provided the perfect point of departure from which to victual the West Indies and/or Biggin's two other North American markets, namely Louisburg and Quebec, in biscuits, flour, bread and rice. New York also played an important role for Biggin: even though he did not actually victual the port, he employed an agent there who was equally active in the logistics behind wider operations in America and the West Indies. This will be seen in more detail in the next chapter with the example of the victualling of the two Caribbean stations.

Biggin's ships, loaded with American produce, then sailed along the New England coast, aided by the prevailing winds and currents, and made their way back to Britain. In this way, Biggin also contributed to the exponentially developing export market from the thirteen colonies to Britain – a market

98 PRO, ADM 111/46, 18 March 1757; ADM 111/53, 10 March, 16 April and 12 May 1762.

Table 6.8 A global overview of markets obtained by John Biggin

	1756	1757	1758	1759	1760	1761	1762
Britain							
East coast ports							
Hull	■						
Leith	■	■	■	■	■	■	■
Lynn	■	■	■				
Newcastle	■	■	■				
Yarmouth							
West coast ports							
Bristol							
Greenock	■	■	■	■	■	■	■
Liverpool							
Milford		■	■	■	■	■	■
Whitehaven							
Ireland							
Belfast/ Carrickfergus		■	■	■	■	■	■
Cork	■	■	■	■	■	■	■
Dublin							
Londonderry							
Waterford							
Intermediary bases							
Oporto			■	■	■	■	■
Lisbon			■	■	■	■	■
Madeira							■
Senegal/Gorée			■	■	■	■	■
West Indies							
Barbados/ Leeward Islands	■	■	■		■	■	■
Jamaica							
Havana							
North America							
Louisburg				■	■	■	■
Nova Scotia							
Quebec						■	■
New York							
South Carolina	■	■	■	■	■	■	■

that was worth £1,543,433 in 1768 according to calculations made by Shepherd and Walton.[99]

Rendered powerful by the geo-strategic positioning of his victualling sources, John Biggin went from strength to strength and continued to conquer markets. By the end of 1762, he had obtained twelve of the twenty-six general markets (see Table 6.8), compared to only seven at the end of 1756.

A closer analysis reveals how John Biggin achieved his breakthrough. Whilst it is true that he lost three contracts (Hull, Lynn and Newcastle on the east coast) during the first three years of the war, he nonetheless began to gain in importance, first by obtaining the victualling of the squadrons operating off the French coast, and then more importantly by succeeding in appropriating most of the foreign markets. Of all twelve contracts in existence at the end of 1762 within the three regions comprising the West Indies, North America and the intermediary bases, John Biggin had succeeded in obtaining no less than eight of them.

Cork played an especially important role in the management of Biggin's victualling. It was from Cork that Biggin supplied the western squadron with livestock and vegetables, and the barrels of pork, beef and butter destined for his other markets also passed through Cork. It is undoubtedly true that his good relations with the cattle breeders and farmers of the area were a major key to his success. At this time, Cork was Ireland's main exporter of cows and pigs and butter, and was consequently the place of choice for buying produce at the most competitive prices.[100]

D. Dickson's study[101] on Cork's merchant community in the eighteenth century also provides an insight into John Biggin's success. In fact, this community of breeders and farmers hardly participated, in terms of their production, in the export market, for the simple reason that, compared to suppliers in Dublin or Belfast, the Cork community possessed very few ships:

> Probably, the main reason why such a large and apparently active centre of trade has achieved so little retrospective notice is that outside of the port, its impact on international trade was disproportionately small. Its merchants owned relatively few ocean-going vessels, they did not have the legal opportunity of participating in the most profitable import trades from the British Colonies, and they worked mainly on a commission basis. Cork in this respect was a Charleston, not a Philadelphia.[102]

99 Shepherd and Walton, *Shipping, Maritime Trade*, 231–2.

100 For this point, refer to the collection of papers given at the conference *Négoce et industrie en France et en Irlande aux XVIII et XIX siècles*, ed. L.M. Cullen and P. Butel (Bordeaux, CNRS, 1980).

101 D. Dickson, 'The Cork merchant community in the eighteenth century: a regional perspective', in *Négoce et industrie en France et en Irlande*, ed. Cullen and Butel, 45–50.

102 Dickson, 'The Cork merchant', 45.

The relationship between Mr John Biggin, ship owner, and this community therefore made the utmost sense. Furthermore, a few of Cork's most important landowners lived in London at the time,[103] which facilitated working with Biggin on a regular basis. Unfortunately, very few economic histories of this region have been published to date – studies that would provide interesting perspectives on this group of owners, breeders, farmers and merchants.

A comparison of John Biggin's prices with those of his competitors throws a less descriptive but more analytical light on his logistical management of victuals. Generally speaking, Biggin was the most competitive when it came to vinegar, which he obtained thanks to his agents in Guernsey and Madeira; Irish produce (beef, pork and butter), unless he was in competition with an Irish merchant; rice, thanks to his links in South Carolina; and peas and porridge oats, thanks to his farms in Britain.

This global infrastructure with its multiple sources seems to have been the key to Biggin's astonishing success and often allowed him to have the edge over local traders, who, whilst they may have offered lower prices for local raw products, could not even begin to compete when it came to imported products, and inevitably quoted higher prices. For example, Biggin could obtain peas, porridge oats and cheese at a low cost from Britain, which allowed him to gain the upper hand over the formidable Auguste Boyd, Ireland's main exporter of beef, and thus win the contract at the invitation to tender bids in Cork. It was also because Biggin was able to combine his foodstuff supplies of Irish origin with those from Britain that he had the edge over local merchants in the intermediate bases, and sometimes even over American merchants for the West Indian market.

There was also the fact that Biggin offered credit facilities together with a guarantee that his foodstuffs would keep for a certain period of time after delivery, which made him a popular choice for the Commissioners of the Victualling Board, despite the fact that these two advantages often resulted in slightly higher prices than his competitors:

> It appears that Mr. Biggin's price of provisions for victualling His Majesty's ships at Oporto turns out 36/100 of a penny a day more than the price charged by Mr. Allen. But the Board thinking the advantage the crown must reap by having provisions warranted and by six months credit instead of paying ready money would be more than the difference in the prices, came to a resolution to agree with Mr. Biggin in case he would agree to take 3 pence per pound for fresh beef.[104]

103 Dickson, 'The Cork merchant', 48.

104 PRO ADM 111/49, 1 August 1759.

It would seem, therefore, that John Biggin's weakest point was his remoteness in relation to North America, not necessarily in terms of sheer distance, but in terms of distance/time, a most important variable in this era of sailing ships. Furthermore, North America benefited from varied food production and a dynamic commercial structure. All these factors, as pointed out by Harold Innis and W. Mackintosh,[105] resulted in John Biggin finding his toughest competitors – Grant, Mauger, Nesbitt and Franks – in North America.

The West Indies proved to be a veritable front line for provisions coming from Europe and from North America. Biggin, for example, succeeded in winning the contract for the victualling of vessels attached to the Barbados and Jamaica stations, but since the latter island was much further from Europe than Barbados in terms of nautical routes, Messrs Nesbitt and Franks from North America won the contract for the victualling of ships in Jamaica from Biggin in 1760.

To conclude, Biggin's supremacy was undoubtedly due to his being the only merchant to have important footholds in Britain, Ireland and North America. These three key positions also allowed him to dominate victualling in the intermediary bases and to establish himself in the West Indies, thanks to the combining of two food sources: Ireland and South Carolina. Thanks also to the Navy expanding its operations in the West Indies and Canada, Biggin continued to win contracts during the latter years of the war (see Table 6.8). Biggin's success was a reflection of the increasingly concentrated direction taken by the victualling of Royal Navy ships in the secondary home ports and the outlying ports and stations.

105 Cited in Farnie, 'Commercial empire of the Atlantic'.

7

An Example of Stations Managed by a Private Contractor in the Service of the Crown: The Caribbean Staging-Posts

Why choose the West Indies as a typical example of the secondary ports and overseas stations by which to illustrate and deepen our understanding of the management of such victualling? First, because apart from the example of the western squadron whose victualling was studied in the chapter focusing on Plymouth, it was in the West Indies that the highest concentration of warships occurred. Second, because of this example's geographical position: the Caribbean (sometimes described as the 'American Mediterranean') provides an excellent opportunity to study the impact of being supplied both from Europe and from America. Finally, because the Caribbean arena had already been researched by the author for a doctoral thesis through a thorough archival analysis covering a wide period from 1689 to 1763.

The period in question was a dynamic one, rich in consequences. Britain found itself in the enviable situation of being increasingly able to meet the needs of the ever growing naval forces that it brought together in this theatre of operations.[1] The victualling process which started out as a series of ad hoc operations took on a more and more effective shape until it became an efficiently functioning system. There were three main stages in this evolution: the setting up of a victualling system in the West Indies under the jurisdiction of the Victualling Board (1689–1715); a period of improvement by experimentation (1715–48); and finally, the era of the contractor (1756–63).

The Establishment of a System of Victualling in the West Indies under the Jurisdiction of the Victualling Board (1689–1715)

During the War of the Grand Alliance, also known as the War of the League of Augsburg, the British Admiralty had to strictly limit the duration of its

1 Buchet, *La lutte*.

campaigns to match the quantity of provisions that could be stocked on board its vessels, i.e. approximately eight months' worth.[2] This was clearly a setback, which Admiral Wheeler tried to correct. Having been given command of a squadron bound for the West Indies in 1693, he drafted a long report for the Admiralty in which he recommended that essential victuals be supplied by the North American colonies. Wheeler could envisage four distinct advantages to this proposal: first, it would better fulfil the squadron's requirements since, he claimed, vessels dispatched from New York or Boston would require under two months to transport the requested supplies to the squadron; second, it would cost less since it was cheaper to purchase victuals from the New World than from England; third, it would extend the campaign time for the expedition; and finally, the proposal would ultimately lead to the development of the infrastructure of Caribbean bases since space in the ships of the squadron which would have been taken up with stocks of victuals could be given over instead to materials necessary for port development.[3] Interesting though they were, Wheeler's suggestions for logistical improvement did not influence the dominant opinion. The Admiral was thus obliged to carry on board his ships, from England, all victuals that would be required, thus reducing his campaign time to under eight months.[4]

Britain had been obliged to maintain a permanent naval presence in the West Indies ever since the Spanish War of Succession, partly in order to protect its flourishing commerce and partly to intercept the treasure fleets which provided its enemies with the wealth they needed to pursue hostilities. In order to render the squadrons as economically viable as possible it was necessary to extend the duration of their operations. Consequently, the Victualling Board had to emerge from the essentially passive role it had adopted in this part of the world, and actively assume the continuous victualling that these changed circumstances now required.

The outlets

Accordingly, as soon as hostilities broke out, the Victualling Board established two outlets, one in Barbados and one in Jamaica, each managed by an Agent.[5]

2 For example, *Calendar of State Papers, America and the West Indies (1689–1692)*, nos. 968 and 1037; *C.S.P., Am. and W.I. (1693–1696)*, no. 167. Victualling Board to W. Blathwayt, dated 4 March 1693.

3 *C.S.P. Domestic, William and Mary, Nov. 1691–Dec. 1692*, ed. W.J. Hardy (London, 1900), 346–52 and 447–8.

4 *C.S.P., Am. and W.I. (1693–1696)*, no. 167.

5 Charles Thomas for Barbados; in Jamaica there was a succession of Agents: Philip Rogers, A. Wilson assisted by Hugh Gaine, Samuel Younghusband, Joseph Gyde and M. Silvester. PRO, ADM 110/2, p. 97; ADM 110/3, pp. 119 and 137; ADM 110/5, pp. 48 and 386.

Aided by their employees, these Agents had two roles.[6] First of all, all victuals sent from Britain or the North American colonies[7] had to be stocked and stored in the best possible conditions – and the Agents were regularly reminded that all ship's captains should apply to them and to no one else to be supplied with extra provisions.[8] Secondly, if, on occasion, the Agents did not possess enough provisions in their store-rooms to satisfy a particular demand, their task was to acquire the missing foodstuffs[9] or, if this was absolutely impossible, to provide the crew with enough cash to buy the victuals themselves elsewhere.[10] Whichever means was used, the success of the operation depended entirely on the Victualling Board's Agent having a supply of money. The Victualling Board ensured that this money would be provided via contracts it drew up with other institutions, and this practice was extended to all Agents operating in ports far from Britain, such as in North America.[11] This was an improvement on earlier methods in which the Agent had to look for a creditor and risked not finding one, or attempted to raise the money by borrowing it at inflated interest rates. This was the unfortunate scenario that had occurred on a number of occasions during the War of the Grand Alliance (the War of the League of Augsburg), when the Agent needed to provide the squadrons with extra supplies.

Thus the Agent based in Jamaica was certain of obtaining loans from the two most important merchants on the island, Gilbert Heathcott and Samuel Kershow,[12] at a fixed interest rate of 8 per cent.[13] The same went for the Agent in Barbados,[14] and both men received orders from the Commissioners '*not to borrow money from other persons in the interests of Her Majesty's service*'.[15] Nonetheless, under certain circumstances, it was permissible for the Agents to place certain expenditures on the Victualling Board's account, as illustrated

6 Their job descriptions were clearly stated in a letter written by the Victualling Board on 8 June 1711 to Mr Lawndes. PRO, ADM 110/5, pp. 228–9.

7 PRO, ADM 110/2, 10 August 1704 , p. 250.

8 PRO, ADM 110/2, 31 January 1705, p. 311; ADM 110/5, 22 July 1710.

9 PRO, ADM 110/2, 3 February 1704, p. 137.

10 PRO, ADM 110/2, 10 December 1703, p. 120.

11 PRO, ADM 110/4, 11 September 1708; ADM 110/3, 21 August 1705.

12 PRO, ADM 110/2, 30 March 1704, p. 170.

13 PRO, ADM 110/5, p. 417. 'An account of the remittances of money for the foreign Service on contracts made by the Commissioners for victualling Her Majesty's Navy during this present war with the rates of exchange and upon what terms and conditions the said remittances have been made.'

14 PRO, ADM 110/3, 21 August 1705.

15 PRO, ADM 110/2, 30 March 1704, p. 170.

by a letter from the Commissioners to the Lord Treasurer in which they ask him to honour their debts.[16]

The methods of liaison

Since provisions from the West Indies themselves were extremely expensive,[17] the Lords of the Admiralty ordered the Commissioners to victual the naval forces in the area with provisions from England by using a system of rotating round trips which operated in the following way: the decision to send victualling ships was not made by the Commissioners but by the Lords of the Admiralty. The Commissioners were then notified by a letter of how much food to send to the West Indies – a calculation based on how many mouths needed to be fed over a certain period of time.[18]

The Victualling Board would then start looking for the necessary victuals, and would also look into the possibilities of chartering merchant ships to transport the goods.[19] It would seem from an examination of the ADM 110 series concerning the War of the Spanish Succession that the Victualling Board did not encounter any problems in procuring ships,[20] especially in relation to those bound for the West Indies.

Charter contracts were drawn up on the basis of a certain price per ton. For example, of thirty-one ships chartered for voyages to Jamaica between 1 January 1701 and 31 December 1707, the price per ton climbed from £4 10s to £6 between 1707 and 1702 and then fell and remained at £3 per ton for the following five years. This decrease in tariff reveals the extent to which charter vessels were readily available, and probably also points to the low risk of loss incurred on these trips.

Most probably the only pressure the charter ship owners felt was running the risk of seeing a voyage lengthened due to delays in loading the merchandise in Britain, or unloading it in the West Indies: this was the '*Plaint of severall merchants who had lett ships to His Majesty's Commissioners of the Victualling for Lisbon and Jamaica*'.[21] This eventually led to the inclusion of protection clauses in the contracts. For example, a contract drawn up in 1702 stipulated that if the ship had not weighed anchor by 1 September 1702, or had not completed unloading within forty days of arriving in Jamaica, the ship owner

16 PRO, ADM 110/3, 26 September 1706, p. 203; ADM 110/5, 23 June 1711, p. 241.

17 PRO, ADM 110/2, 29 April 1703; ADM 110/3, 22 April 1706; ADM 110/5, 21 April 1712.

18 PRO, ADM 110/2, 13 February 1705; ADM 110/3, 15 April and 31 December 1707.

19 PRO, ADM 110/2, 13 April and 5 May 1704; ADM 110/3, 15 April 1706; ADM 110/4, 27 October 1709.

20 PRO, ADM 110/3, 31 December 1707.

21 PRO, ADM 110/2, 5 May 1704.

was entitled to claim compensation.[22] Once the merchant ships were almost completely loaded, the Commissioners would inform the Lords of the Admiralty of the extent to which they were managing to keep to schedule and would specify the date at which they planned to be at The Nore – the usual point of departure for voyages to the West Indies. The merchant ships would then be escorted from The Nore by warships.[23]

These deliveries of victuals from England occurred on an irregular basis. As Table 7.1 shows, it seems that the rotating round trips were decided upon empirically rather than being based upon a regular service with timed intervals.

The development of commercial links with North America and the effect this had on the Navy's victualling practices

The development of the system of importation from North America would prove to have a beneficial effect both on the victualling of the fleet in the West Indies and in modifying the structure and management of this victualling. Already by 1703, Whetstone's fleet had been amply victualled from the North American continent and indeed a Crown official in Philadelphia, acting on instructions from London, offered to send bread and other perishables on a regular basis on the condition that the vice-admiral supply the necessary ships.[24] In fact, this was nothing compared to what would transpire between 1705 and 1713.

Despite the irregular pattern of the round trips, their frequency multiplied and consequently the foodstuffs imported into Jamaica, mainly from New York and Virginia, became so important that the island found itself to be a crossroads of sorts for all commerce with Spanish possessions.[25] By 1708, English possessions in the West Indies were so dependent on victuals from the North American colonies that the Council of Trade and Plantations requested the Queen to expel the French from the continent so as to ensure that they would not be able to cause any harm to these commercial connections, which were deemed to be 'so necessary'.[26] By 1709, American

22 PRO, ADM 110/2, 5 May 1704. Owners were anxious to get their ships out of the Caribbean before the start of the hurricane season for insurance and safety purposes.

23 PRO, ADM 110/2, 13 and 27 February 1705; ADM 110/3, 22 April 1706.

24 *C.S.P., Am. and W.I. 1702–03*, no. 645. Quary to the Admiral in command of the fleet based in Jamaica.

25 *C.S.P., Am. and W.I. 1708–09*, no. 779. The Council of Trade and Plantations to the Earl of Sunderland (18 Oct. 1709; no. 382), The Council of Trade and Plantations to Governor Handasyd.

26 *C.S.P., Am. and W.I. 1708–09*, no. 221. The Council of Trade and Plantations to Her Majesty the Queen (1 Dec. 1708).

Table 7.1 Convoys of chartered ships bearing victuals from England to Jamaica

Date of decision	Number of ships	Provisions	References
July 1702	9 ships:		
	Duke of Gloucester		PRO, ADM 110/2, p. 193
	Bonadventure		PRO, ADM 110/4 (at
	Speedwell		the end of this register,
	Chamberlain		unpaginated, 'An account of
	Thomas and Elisabeth	?	the extraordinary …')
	Rainbow		
	Expedition		
	Florence Freigh		
	Hawke		
August 1702	2 ships		
	Britannia	?	PRO, ADM 110/4 (at
	Diana		the end of this register, unpaginated, 'An account of the extraordinary …')
January 1703	4 ships	?	*C.S.P., Am. And W.I. (1702–1703)*, no. 1389i
13 April 1704	1 ship		
	Norwich	Irish meat Bread	PRO, ADM 110/4 (at the end of this register, unpaginated, 'An account of the extraordinary …')
November 1704	5 ships		
	Robert		PRO, ADM 110/4 (at
	George	?	the end of this register,
	Anna		unpaginated, 'An account of
	Mary and Martha		the extraordinary …')
	Richard and Sara		
13 February 1705	6 ships	4 months' worth of provisions 'at whole allowance'	PRO, ADM 110/2, pp. 318, 324
15 April 1706	3 ships		
	Hunter	3 months' worth	PRO, ADM 110/3, pp. 114,
	Francis and Elisabeth	of provisions for	120
	Herbert and Galley	2,000 men 'at whole allowance'	PRO, ADM 110/4 (at the end of this register, unpaginated, 'An account of the extraordinary …')
23 October 1706	6 ships		PRO, ADM 110/3, pp.
	Amazon	6 months' worth	213–14, 220
	Robert and Francis	of provisions for	PRO, ADM 110/4 (at
	Marlborough	2,000 men	the end of this register,
	Lawra		unpaginated, 'An account of
	George		the extraordinary …')
	Gold Fright		
29 January 1707	3 ships		
	Jersey	4 months' worth	PRO, ADM 110/3, pp. 269,
	Crown	of provisions	273
	Maidstone		

9 June 1707	1 ship *Anna Frigot*	 ?	 PRO, ADM 110/3, p. 317 PRO, ADM 110/4 (at the end of this register, unpaginated, 'An account of the extraordinary ...')
31 December 1707	*Colchester* ?	6 months' worth of provisions for 3,000 men 'at whole allowance'	PRO, ADM 110/3, p. 425 PRO, ADM 110/4 (at the end of this register, unpaginated, 'An account of the extraordinary ...')
13 April 1708	6 ships *Rutch* *Colchester* *Aleppo Factor* *John and Thomas* *Mary and Kath* *Guilford*	 6 months' worth of provisions for 3,000 men 'at whole allowance'	 PRO, ADM 110/4, p. 132
8 February 1709	2 ships *Chatham* ?	 ?	 PRO, ADM 110/4, p. 132
27 October 1709	4 ships	6 months' worth of provisions 'at whole allowance'	PRO, ADM 110/4, p. 298
29 March 1711	8 ships *Defyance* *Burlington* *Jamaïca Sloop* *Tryall Sloop* *Salisbury* *Panther* *Jolly* *Bedford*	 ?	 PRO, ADM 110/5, p. 180

Number of Merchant Ships which arrived in Jamaica between
25 March 1709 and 29 September 1711
from England: 104
from the Colonies: 207

These imported products were mainly beef, pork, fish, flour, butter, cheese, beer and wine. (*C.S.P., Am. and W.I. July 1711–June 1712*, no. 267 ii)

production of cereals was so abundant that London was considering selling its surpluses to the Portuguese.[27]

In Barbados and the Leeward Islands, commercial ties with the northern colonies appeared to be so solid that the Victualling Board decided, on 24 September 1705, to entrust all victualling of Navy ships to a single contractor, a Mr Collins, who was firmly established in North America.[28] His contract lasted until 1710, when he was replaced by Thomas Pindar, whose contract was based on the going price of 14 pence per daily ration per head.[29]

In this way, the buying of victuals and other supplies from regional markets, plus their storage, was entirely dependent on the initiative being taken by the local contractor. This dramatic change in the management of victualling, made possible only by regular communications with North America, offered two advantages to the Admiralty: first, the competitive prices and excellent quality of products from the northern colonies. According to Colonel Quary, the Victualling Board's official representative in Philadelphia, the price of bread and flour was much less expensive than in England, and furthermore, '*one could not find bread and flour of a better quality anywhere else, and the same goes for rice, beans and pork*'.[30] The only exception was salt beef, which could not be preserved for more than six months. The second advantage was that the pressure was taken off the Victualling Board somewhat, especially since it was hard-pressed to keep up with important naval operations being carried out in Europe. For all these reasons, the Victualling Board had again entrusted its operations to Mr Collins when, in 1708, the Queen decided to come to the aid of the inhabitants of the Leeward Islands. This time, the contractor was asked to deliver 2,400 barrels of beef at £3 2s 3d per 200 lb barrel, and 400 barrels of pork at £5 per 200 lb barrel.[31] Collins, followed by Thomas Pindar who succeeded him in October 1710,[32] provided such a good service in victualling the warships in Barbados and the Leeward Islands, that the Victualling Board complained to the Lords of the Admiralty on a number of occasions that food consumption in these ports was excessive.[33]

27 *C.S.P., Am. and W.I. 1708–09*, no. 779. The Council of Trade and Plantations to the Earl of Sunderland (18 Oct. 1709).

28 PRO, ADM 110/3, 24 September 1705.

29 PRO, ADM 110/5, 14 February 1711.

30 *C.S.P., Am. and W.I. 1702–03*, no. 645. Quary to the Earl of Nottingham.

31 PRO, ADM 110/4, 1 January 1708; ADM 110/5, 11 September 1710.

32 PRO, ADM 110/5, 14 February 1711.

33 PRO, ADM 110/6, 10 June 1713. This document reveals that over-consumption of victuals had already been noted on numerous occasions. See also PRO, ADM 110/5, 5 August 1712; ADM 110/6, 27 September 1713.

The Victualling Board itself continued to supply warships in Jamaica with their victuals throughout the duration of the war, either by convoys sent from England or by arranging to buy local goods thanks to the financial agreements drawn up with local merchants.

But why was there such a difference between the modes of victualling in Barbados and in Jamaica? A letter from the Victualling Board to the Lords of the Admiralty in July 1706 offers the principal motives.[34] The letter states that in 1706, a merchant called Sir Geoffrey Jefferys who was already responsible for the victualling of the Navy in New York suggested to the Board that he also take on the victualling of Her Majesty's fleet in Jamaica. The commissioners thought it best to turn down this offer, partly due to the fact that there were many more vessels permanently stationed in Jamaica than in Barbados, and they were fearful of delays in the victualling of these ships. The main reason, however, was financial. In fact there was a disagreement over the type of service offered, which would have resulted in some perceptible differences in price.[35] Jefferys was asking to be paid on the basis of victuals that he would deliver to the Victualling Board's Agent in Jamaica, whilst the Commissioners only wanted to pay him for victuals supplied to warships. In the first set of circumstances, the supplier would simply send the requested victuals to New York and it was the Victualling Board which covered the cost of transport, storage in Jamaica and the eventual loss of some perishables due to either poor preserving methods or simply an excess of the product. In the second set of circumstances, it was the supplier who covered all of these possible costs, and was paid on the basis of actual victuals delivered to the ships – in other words, the merchant had to cover any eventual losses himself. This option, moreover, gave the Victualling Board the possibility of no longer having to maintain personnel in Jamaica (as was the case in Barbados) – a financially attractive proposition.

As a result of the large number of vessels stationed in Jamaica and the ensuing fluctuations in the number of ships to be victualled, neither Jefferys nor any other merchant was willing to subscribe to the Victualling Board's conditions, thus forcing the Commissioners to continue victualling the port themselves, unlike the situation in Barbados.

However, the development of commercial ties with North America did have some positive effects on the victualling of expeditions sent to Jamaica. Basically, whenever food supplies ran short due to delays in the arrival of the convoys or because supplies had not been properly conserved – and this happened on a fairly regular basis – the provisions could be bought locally

34 PRO, ADM 110/3, 31 July 1706.

35 See J. Jeffery's proposals in PRO, ADM 110/3, 31 July 1706.

by a Victualling Board agent or even by the crew, since the men were given an allowance to buy food, called '*short allowance*', to be used precisely in such circumstances.[36]

Thus, despite inevitable shortages in supplies from England, it was still possible to maintain acceptable levels of victualling throughout the War of the Spanish Succession, thanks to the sheer quantity of products from North America, together with certain financial arrangements drawn up with the local colonists. In sum, the economic development of the North American colonies provided the basis for a new type of logistical organisation with a broader intercontinental perspective.

Empirically Driven Improvements (1715–48)

The war having ended, Britain nonetheless decided to maintain a naval presence in the West Indies. From this moment on, the Victualling Board, which had been directly responsible for supplying those squadrons based in Jamaica throughout the War of the Spanish Succession, handed over the task to private enterprise – which had been the case in Barbados since 1705. This type of organisation was to last in Jamaica until 1741, when the Victualling Board once again decided to directly manage the victualling of vessels stationed at that island.

What provoked this turnaround? What were the advantages and the inconveniences of the two systems? The answers to these questions can be determined by a thorough examination of the logistics of provisioning Vice-Admiral Vernon's expedition, since it was during this operation that the decision to revert to the old system was taken.

The victualling of Vice-Admiral Vernon's expedition and the relinquishment in Jamaica of a privately operated victualling system

In 1739, the year in which Vernon was sent to the West Indies, the Admiralty had enormous faith in the private system and in their suppliers, a company

36 PRO, ADM 110/3, October 1706: 'A general account of the money necessary for payment of bills of exchange and necessary money for short allowance in preference of other services of the Victualling'. PRO, ADM 110/3, 5 September 1707, Victualling Board to the Admiralty. PRO, ADM 110/4, pp. 110, 271: 'List of bills of exchange drawn for the service of victualling which remains unpaid'. PRO, ADM 110/4, p. 357: Victualling Board to Admiralty (4 Feb. 1710); PRO, ADM 110/5, p. 45: 'Accounts of bills of exchange drawn for the service of victualling with what is wanting for necessary and extra-necessary money'. PRO, ADM 110/5, Victualling Board to Admiralty (23 June 1711). *C.S.P., Am. and W.I. 1711–12*, no. 267, Hamilton to the Council of Trade and Plantations (19 Jan. 1712). The following letter seems most to illustrate the relative ease of victualling the West Indies: PRO ADM 110/5, Victualling Board to Admiralty (21 April 1712)

called Mason and Simpson, which had been looking after the needs of reduced naval forces based in Jamaica for the past ten years. So great indeed was the Admiralty's confidence, that it failed to inform its contractor of the arrival of Vernon's large squadron until after it had left England. This resulted in a shortage of peas, which led the Mason and Simpson Company to write a sharp letter to the Commissioners of the Victualling Board:

> If there is any intention or reasonable probability to send a further number of ships to Jamaica, we must desire and insist on a timely intimation of it. For as the island produces but a few species for the victualling of the ships, and in small quantity, we hope we cannot be thought blameable should our stores not be capable of supplying them in case of their coming thither, unless timely notice be given us what provisions to provide for them.[37]

Slowly but surely, the Admiralty became increasingly disillusioned with this system, especially in the context of a large-scale expedition. Then, in November 1739, three months after Vernon's departure, the contractor requested a supplementary £10,000 from the Admiralty in order to be able to continue victualling the squadron, 'otherwise we shall be unavoidably necessitated to put an end thereto ....'[38] The reasons given by Mason and Simpson were that, first, they felt it necessary to have at least three to four months of provisions in storage to ensure that the victualling needs of 4,000 men would be safely covered, 'which will require so large a sum of money that the price of our present contract will no way answer the value of the interest of money to be employed'.[39] Second, due to the war, they predicted being charged higher interest rates by their creditors whilst at the same time the cost of beef had already increased by 10 per cent, mutton had climbed even higher and, to make matters worse, there was a total dearth of beans on the market – a staple item in a sailor's diet – due to the season's catastrophic harvest.

Consequently, Mason and Simpson could only continue to victual the squadron by modifying the contents of daily rations as set out by the Admiralty. Flour, fish, oil, vinegar and rice were partially replaced by substitutes such as bread, beef and peas. This clearly led to tense moments during the negotiations in which both parties had to agree on a price regarding this modification of the contract.[40] The Commissioners advised the Lords of the Admiralty to grant the requested sum of money, not so much from a sense of

37 B.M. Ranft, *The Vernon Papers* (London, Navy Records Society, 1958), vol. 99, no. 259, Mason and Simpson to the Victualling Board, 20 November 1739.

38 Ranft, *The Vernon Papers*, no. 259.

39 Ranft, *The Vernon Papers*, no. 259.

40 PRO, ADM 110/13, 19 March 1741.

conviction, but simply to gain time so that the men would not suffer while a reorganisation of the system took place.[41]

A letter written by Vernon in September 1740 was to hammer another nail in the coffin of the private system.[42] In fact, Mason and Simpson's contract had been maintained, indeed re-launched, on a certain number of points on 24 April 1740, when they were entrusted with the victualling of 6,000 soldiers for a period of one year – soldiers the Admiralty was proposing to send under the command of Rear Admiral Ogle and General Cathcart. Upon hearing this, Vice-Admiral Vernon, who was already complaining about a severe shortage of beans, did not hesitate to voice his discontent:

> The times past have been times of peace and the numbers of men to be victualled here by that contract is in proportion very few to the present. And having observed that they principally rely on supplies from the Northern Colonies for making it good, I think it my duty to represent to you that I am greatly apprehensive if they depend on that, His Majesty's Force will run the hazard of starving here and for these reasons. Viz.
>
> That now their demands is likely to be so increased, not only by the large increase of the squadron, but by their having contracted likewise to victual all the land forces coming here. It is much to be doubted that the growth of that country can't supply. And if as in fact it will be so, the demands from these parts, both on account of the Spaniards and French should increase too. I believe it will be found by experience that they choose to supply them even preferably to our own colonies, having their payments from them either in money, or commodities they can get at lower price with the French than us. And they will supply them even in a war by the way of Eustatius and Curaçao, and now they do likewise by Port François on the north side of Hispaniola. And I believe that till the market is glutted they don't come to Jamaica at all. And therefore if these supplies are to be depended on from thence, as I conjecture from those words of the contract that says. They shall be the best provisions of each kind that Jamaica does afford, I may venture to say without the spirit of prophecy, there will certainly be a disappointment in the contract. As I am persuaded gentlemen that you will agree with me that it is equally incumbent on you to inquire into the practicability of executing the contract as well as having made it.[43]

As a result therefore of both rising costs and Vernon's fears, the Lords of the Admiralty decided to consult with the Victualling Board on 21 October 1740 in order to debate whether or not the private system should be abolished,

41 Ranft, *The Vernon Papers*, no. 258. Victualling Board to the Admiralty, 23 November 1739.

42 Ranft, *The Vernon Papers*, no. 261. Vernon to the Victualling Board, 4 September 1740.

43 PRO, ADM 110/13.

especially in light of an expeditionary force of 20,000 men which was soon going to be deployed.[44]

In its reply dated 27 October, the Victualling Board recommended that the private system be dropped, and offered to be directly responsible for the victualling of the expedition. There were two reasons behind this decision. First, the Victualling Board doubted that the contractor would be able to assemble the enormous quantities of provisions necessary for the victualling of 20,000 men on his own authority. The second reason was a financial one. The cost of victualling 20,000 men for one year had been estimated at £300,000. If the private system were to be maintained, then it would have been customary for the Victualling Board to pay one-third of the sum to Mason and Simpson immediately, and to pay another third within four months. Faced with such a great financial burden, it seemed more prudent for the Victualling Board to take over the operation since the initial outlay would then be avoided. Indeed, it was customary for the Board to spread their costs out progressively by taking advantage of credits provided by their suppliers (see Chapter 3).[45]

The results obtained by the management of provisions in Jamaica by the Victualling Board alone

The Victualling Board thus took over the victualling of the naval forces based in Jamaica, just as it had done during the preceding war. However, the special circumstances in which operations were carried out between 1741 and 1748 did lead the naval administration to modify certain aspects of this victualling, and on occasion to adopt different methods. These special circumstances were essentially created by the higher numbers of men involved. Most of the time, 7,000 men were stationed on the island, a figure that on occasion rose to 20,000 men and never dropped below 3,940 men. Compare this to statistics for the War of the Spanish Succession, in which there were on average 2,000 men stationed on the island, never rising beyond 4,000 men.

Thus the Victualling Board once again established an outpost in Jamaica, this time more heavily manned than during the previous war. The team consisted of fifteen employees: two Agents, Messrs Gray and Maynard,[46] six clerks, one master cooper, four coopers and two storekeepers.[47] As in the past, the team's task was to stock the provisions, keep them in good preservation,

44 Ranft, *The Vernon Papers*, no. 262, Victualling Board to the Admiralty, 27 October 1740.

45 Ranft, *The Vernon Papers*, no. 262, Victualling Board to the Admiralty, 27 October 1740.

46 PRO, ADM 110/13, 7 April 1742.

47 PRO, ADM 110/13, 7 April 1742 (List of expenses for payment of Victualling Board personnel).

and take the initiative in providing vessels with victuals procured by themselves in the event of stocks running out. Unlike the preceding war, it seems however that no agreement was ever concluded to set up financial arrangements between the Victualling Board Agents and the local colonists, which would have enabled the Agents to acquire the money necessary to procure victuals if and when necessary. On the other hand, the rather substantial sum of £10,000 was sent directly to the Agents from the London office.[48] Moreover, orders issued by the Privy Council on 24 December 1740 forbad Jamaica's Assembly from levying taxes on any imported provisions destined for the Navy.[49]

Once again, just as in the War of the Spanish Succession, provisions were transported to Jamaica using a system of rotating round trips from England. As soon as the decision to send a convoy was taken, the Victualling Board would send word to its storehouses in Gravesend to prepare the provisions. At the same time, it would charter some merchant ships in order to transport the provisions to Spithead via The Nore.[50] This alone could take up to six weeks.[51] From Spithead, the chartered merchant ships, escorted by warships,[52] would set sail for the West Indies, which they would reach about two months later. It was thus necessary to anticipate the needs of the naval forces in Jamaica three and a half to four months in advance.[53]

This section will conclude by briefly examining the victualling of wine. At the time when Mason and Simpson were still in charge of victualling the expedition led by Vice-Admiral Vernon, a contract was drawn up, in March 1740, with Sir George Champion for the supply of 600 pipes of wine from Santa Cruz in Tenerife.[54] Now that the Victualling Board had regained control of sending provisions to Jamaica, they also drew up a similar contract, on 7 February 1741, with a certain Mr Reynolds. However, the execution of this contract was to provoke a fiscal dispute since Reynolds did not deliver the wine to the Victualling Board's storerooms in Jamaica, but to his own warehouse. This meant that he was subject to an import duty of 18 per cent per gallon on his wine. Reynolds refused to pay this tax, quoting the Privy Council's order of 24 December 1740 exempting all Navy provisions imports from taxation, and claiming that he was being unfairly treated and that the

48 PRO, ADM 110/13, 7 April 1742.

49 PRO, ADM 110/13, 16 June 1742.

50 PRO, ADM 110/13, 25 August, 31 October, 27 December 1743; 14 March 1744.

51 PRO, ADM 110/14, 29 March 1745.

52 PRO, ADM 110/13, 25 August 1742, 4 March 1743, 8 February 1744; ADM 110/14, 25 May and 31 July 1744, 29 March 1745.

53 PRO, ADM 110/13, 17 February 1744.

54 PRO, ADM 110/13, 17 December 1742.

order was being interpreted too restrictively. His vessel was actually seized by the island's authorities until the Admiralty intervened on his behalf.[55]

In conclusion, it would appear that the victualling of soldiers and seamen throughout this seven-year period, with provisions sent from England, was generally successful, despite the large numbers of men involved. Consequently, there was less need for the safety margin of American provisions, bought directly by the Victualling Board's Agents in Jamaica. In fact by 1744, there was such an abundance of provisions being sent from England that the Agents in Jamaica, Messrs Gray and Maynard, saw fit to sell some of them.[56] This situation was in direct contrast to events which occurred during the War of the Spanish Succession, in which the victualling organisation broke down even though there were fewer mouths to feed.

Table 7.2 summarises all dispatches from England to Jamaica in the 1742–8 period, with a view to better understanding the successes of the system, but also its limits.

The table clearly shows the regularity and frequency with which provisions were sent to Jamaica – on average, every four months. This is in marked contrast to the irregular and infrequent round trips that occurred from England to Jamaica during the preceding war. Also, the convoys in the preceding war usually carried about six months' worth of provisions, whereas Table 7.2 shows that the convoys now carried enough food to victual the men for about four months. This particular difference had nothing to do with the fact that there were more men to feed, but rather with the Victualling Board's desire to minimise food loss due to poorly preserved provisions. Thus the triennial rhythm that the Commissioners managed to achieve – despite a relative shortage of warships – ensured fairly complete victualling from England to the naval bases in Jamaica. This apparent success is corroborated by an examination of the correspondence written by the squadron commanders (to be found in the ADM 1 series). It would appear that the system broke down on only three occasions, but even then the naval forces did not suffer, since the Victualling Board's Agents in Jamaica and/or the Commissioners in London contacted the Agent in New York who arranged for provisions to be sent from the New World.[57]

The private system maintained in Barbados: causes and effects

Throughout the war, the victualling of naval forces in the Leeward Islands was carried out from the staging-post in Barbados and was controlled by private

55 PRO, ADM 110/13, 16 June 1742.

56 PRO, ADM 110/14, 25 January 1745.

57 PRO, ADM 1/233, f. 745. Davers to the Admiralty, 9 March 1746; ADM 1/233, f. 812. Davers to the Admiralty, 18 June 1746; PRO, ADM 110/15, 25 February 1747.

Table 7.2 Chartered convoys from England to Jamaica

Date of decision	Number of ships	Supplies on board	References
2 April 1742	5 ships were being prepared and were to be sent by the *Adventure*	3–4 months' worth of provisions for 8,000 sailors and 4,000 soldiers	PRO, ADM 110/13, pp. 10–11, 18, 26–8.
25 August 1742 The ships were supposed to set sail from Gravesend before 25 September	6 ships	Provisions for 7,000 men	PRO, ADM 110/13, pp. 114 and 132.
February 1743	3 ships escorted by the *Oxford*	?	PRO, ADM 110/13, p. 311.
4 March 1743 Departure planned for 7 March from Gravesend	1 ship: the *Great Britain* escorted by the *Severn*	20 weeks' worth of provisions for 5,000 men in bread and peas because the Victualling Board had just learned that these provisions were missing	PRO, ADM 110/13, pp. 222, 228 and 311.
12 July 1743 Convoy to depart in September	4 ships	12 weeks of bread 6 weeks of raisins 8 weeks of pork 4 weeks of peas 4 weeks of olive oil 12 weeks of vinegar for 6,000 men This cargo had to victual the fleet until February	PRO, ADM 110/13, pp. 311, 315 and 472.
31 October 1743 At the end of December, the ships were at Spithead. They left this port bound for the West Indies at the end of January	5 ships escorted by the *Biddeford* and the *Kinsale*	4 months' worth of provisions for 7,000 men. This convoy had to victual the men until the end of July	PRO, ADM 110/13, pp. 387 and 390. The detailed contents of this shipment are laid out in a letter from the Victualling Board dated 24 November 1743: PRO, ADM 110/13, p. 409. See also: PRO, ADM 110/13, p. 435, ADM 110/14, pp. 4–5.
20 February 1744 At the end of July the ships were at Spithead. They would leave in mid-November with the squadron led by Davers	6 ships then 5 *Londavery* *Rhodes* *Pembroke* *Carteret* *Bridgedown* Escorted by Davers's squadron, but the *Carteret* and the *Bridgedown* would lose the convoy and return	4 months' worth of provisions for 7,000 men ('at whole allowance')	PRO, ADM 2/61, pp. 55, 487 and 490. PRO, ADM 110/14, pp. 4, 25, 106–7.

7 August 1744 The ships were to set sail from Gravesend before 10 September	6 ships *Snapper* *Price Frigate* *Lovely Mary* *William* *Francis* *Boston*	4 months' worth of provisions for 7,000 men. The contents of this shipment together with the previous one were supposed to feed 6,670 men at whole allowance until July	PRO, ADM 2/ 62, p. 56. PRO, ADM 110/14, pp. 60, 75 and 123.
	Also escorted, it seems, by Davers's squadron		
15 January 1745	2 ships *Harrington* *Heathcote* Escorted by the *Suffolk* and the *Dreadnought*	These two ships were big enough to carry 4 months' worth of provisions for 6,670 men. This shipment was supposed to victual the men until Christmas	PRO, ADM 110/14, pp. 134 and 188.
16 May 1745 On 12 November, due to a lack of escort ships, the convoy was still at Spithead. Consequently, it was necessary to replace some of the provisions which had gone bad. The convoy finally left on 25 January 1746	4 ships at first, then 5 *Lucrétia* *Sally* *Bedford* *Shirley* *Theophilia*	4 months' worth of provisions for 6,000 men. (The delay had certain limited consequences in that the numbers of conscripts for the fleet stationed in the West Indies were reduced following the departure of Rear Admiral Ogle) (cf. PRO, ADM.110, no. 14, p. 385)	PRO, ADM 110/14, pp. 220–1, 379 and 385.
	The *Lucrétia* and the *Bedford* ended up not in Jamaica but in Gibraltar		PRO, ADM 110/14, p. 390.
2 June 1746	1 ship *Harrington* (600 tons)	Bread was the main shipment due to pressing needs	PRO, ADM 110/14, p. 451.
5 February 1747	1 ship *Carteret* (600 tons)	'300 and 400 tons which will be sufficient to carry on the victual (together with what are in stores and aboard ships) in all species for 3490 men to about the 21 August next' (almost 4 months)	PRO, ADM 110/15, pp. 22, 26 and 33.
16 May 1747 Was waiting for its escort at Spithead	1 ship *Richmond* (500 tons) Escorted by the *Warwick*	This shipment was supposed to ensure the men's victualling until about Christmas (about 4 months away)	PRO, ADM 110/15, pp. 74, 85 and 103.
16 November 1747	1 ship *Baltimore* (600 tons)	14 weeks' worth of provisions for 4,000 men	PRO, ADM 110/15, p. 181.
22 January 1748	1 ship *Carteret*	?	PRO, ADM 110/15, p. 223.

merchants under contract to the Victualling Board. Up until 1744–5, victualling was in the hands of Henry Lascelles,[58] an important London merchant, who was succeeded by Mason and Simpson who had their contract renewed on 8 February 1747.[59] The private system was maintained in Barbados (despite the fact that the Victualling Board had taken over the victualling of Jamaica as of 1741) quite simply because there was never any doubt as to the contractor's ability to furnish the required quantities of provisions, due to the lower numbers of men stationed in Barbados. In fact, until 1743, the number of vessels putting in at the island was quite insignificant. However this situation was to change in 1743 when the Leeward Islands became a new autonomous naval station, with its own commander in chief (as opposed to being under the commander in chief posted in Jamaica). At this point, the number of ships stationed at Barbados increased, but consistently remained lower than the number in Jamaica.

At this time, the Victualling Board was content simply to insist that if and when the Admiralty decided to send extra ships, the local suppliers be advised of the decision as soon as possible.[60] This precaution resulted in the following incident: towards the end of 1742, Chaloner Ogle instructed Admiral Knowles to take possession of La Guayra and thus to take command of ten ships from the naval forces in Jamaica. However, Barbados was not warned that the ships would be putting in there for victualling purposes. Consequently Henry Lascelles, the contractor in Barbados, finding himself in short supply, was obliged to offer different, substitute provisions that were of an inferior quality. The Victualling Board would later demand that Lascelles reimburse them.[61]

Before examining the efficiency of the private system of victualling in the Leeward Islands, it is worth noting that the types of provisions allocated to the two stations in the Caribbean differed on one point. Crews putting in at Jamaica were supposed to receive fresh meat once or twice a week, whilst there was no provision for fresh meat in the Leeward Islands, since the climate was considered to be healthier in this region:

> We have examined into the books of our office for upwards of forty years past and don't find that there has been any contract made by supplying any of Her Majesty's Ships there (Antigua) or at any of the Leeward Islands with fresh beef. The reason of which we are of the opinion must have been owing to the

58 PRO, ADM 110/14, 7 November 1746.

59 PRO, ADM 2/70, Lords of the Admiralty to the Victualling Board, 14 September 1747; ADM 110/15, 11 September 1747.

60 PRO, ADM 110/14, 25 May 1744.

61 PRO, ADM 2/61, Lords of the Admiralty to the Victualling Board; ADM 110/14, 18 July 1744.

> healthiness and temperature of those islands in preference to Jamaica especially of the ships kept much at sea.[62]

Apart from the incident mentioned above, no other recriminations on the part of the commanding officers or the Victualling Board against the contractors seem to appear in the documents, which would imply that the private victualling system must have functioned adequately.

On a financial level, there was, however, one important dispute. Mason and Simpson were of the opinion that a tax of 4.5 per cent that was levied on their importation of rum by the authorities in Barbados (based on an Act voted by the island's Assembly in 1663) ran contrary to the contract they had signed with the Victualling Board. The Commissioners were obviously in agreement with the contractors because they advised the Lords of the Admiralty to reimburse Mason and Simpson and to request that the Treasury annul the Act forthwith, or run the risk that prices negotiated in the contract no longer be valid.[63]

The Commissioners' advice appears to have been heeded, and indeed, three months later on 8 February 1747, Mason and Simpson signed a new contract in which they accepted conditions which were far more advantageous to the Crown, in terms of price per ration and in terms of a guarantee that provisions would remain edible up to four months following delivery.[64]

Of the two systems that existed during these wartime years – one under the direct jurisdiction of the state, the other provided by private groups – which was the most efficient? The answer might be found in the letter reproduced below, in which the Commissioners of the Victualling Board, encouraged by their firsthand experience, pressed the Lords of the Admiralty to reintroduce the competitive system by which to choose contractors in Jamaica. The Commissioners claimed that this would result in a saving of 14 per cent (*'one seventh part of the expense'*):

> We pray you will please to acquaint their Lords that from a computation that we have caused to be made of the expense of victualling His Majesty's Ships at Jamaica for the year 1745 and 1746 and by experience we have lately had in making a new contract for the supply of those at Barbados and Leeward Islands, we are of the opinion that by making a contract for the victualling at Jamaica also, there is a prospect of saving about one seventh part of the expense or near two pence a man a day and that there is some reason to think the men may be saved with the dry species of provisions something fresher by a contract than they are at present, and therefore if their Lords shall think it expedient

62 PRO, ADM 110/13, 12 November 1742.

63 PRO, ADM 110/13, 7 November 1746.

64 PRO, ADM 2/70, Lords of the Admiralty to the Victualling Board, 14 September 1747; ADM 110/15, 11 September 1747.

> to put H.M. Ships there under that method of victualling, we desire you will please to signify their directions for our receiving tender for a contract without which it cannot certainly be known what the price will be.

We think it necessary to add that this contract being a great undertaking at the season for killing flesh in Ireland from whence we apprehend it will be served being over for this year we believe no person can be provided to commerce upon it till about Christmas next, so that if we receive their Lords orders to receive proposal about the end of June, we apprehend it will be a proper time and people will know better what price to tender at and we shall be better able to judge of the same of their Lords will have the longer time for considering and resolving thereupon.[65]

The Age of the Contractor (1756–63)

During the Seven Years War, and contrary to preceding wars, Jamaica and the Leeward Islands were both victualled by private contractors, despite the greater number of mouths to feed – namely about 6,000 men in each of the two zones. The figure rose to 30,000 men in 1762, during operations against Havana. How did this system progressively improve, and how did it manage to provide for such large numbers?

Improvements made to the private system

The victualling of the two yards was carried out, as seen in Chapter 6, by separate suppliers according to contracts with only a few differences. For a period of thirteen years, from 1748 to 1760, Mr Auguste Boyd was in charge of the victualling of warships putting in at Jamaica, having signed a contract on 29 June 1748 which was renewed on 6 June 1755. He was succeeded in May 1760 and until the end of the war by the Nesbitt and Franks Company[66] which was well established in North America. The suppliers to the Leeward Islands were successively J. Biggin (from 13 August 1756 to 1 May 1758),[67] S. Fouchet (from 1 May 1758 to 7 March 1760)[68] and then J. Biggin again from March 1760 to the end of the war.[69]

The posts the contractors held in Jamaica and the Leeward Islands differed on three points. First of all, the Victualling Board had taken upon itself to

65 PRO, ADM 110/15, 8 January 1748.

66 PRO, ADM 110/20, 18 January 1762.

67 PRO, ADM 110/18, 23 March 1757; ADM 110/19, 31 May 1759.

68 PRO, ADM 110/20, 31 March and 21 December 1761.

69 PRO, ADM 110/20, 21 December 1761, 10 March and 16 April 1762.

enlarge the warehouses at New Greenwich in Jamaica at the time when it controlled victualling there. Consequently, the Commissioners decided not to loan but to rent this space to the suppliers on the basis of an annual rent of £100 from June 1748 through to October 1756,[70] and henceforth increasing the rent to £300 per annum.[71] Secondly, in contrast to Jamaica, contractors in the Leeward Islands had to guarantee the good preservation of their produce for a certain length of time after delivery[72] (just as in the case of Mason and Simpson, according to their contract of 8 February 1747).[73] Thirdly, in Jamaica, the suppliers were only ever asked to deliver their produce to the warehouses in New Greenwich, whilst in the case of the Leeward Islands, contractors were obliged to deliver to Barbados and also to the new base at English Harbour, Antigua.[74]

Apart from these three minor differences between the two Caribbean stations, there was yet another one, of much greater importance since it had a consequence in relation to the origins of the provisions. Even though there is a distinct lack of archival evidence, there is no doubt that Jamaica was far more dependent on provisions from North America than either Barbados or the Leeward Islands, in which there was a strong European presence amongst the suppliers. It appears that half of the Leeward Islands' victuals originated in Ireland, from which they received beef, pork and butter,[75] whilst the other half came from the North American colonies which provided bread, flour, rice and oat flakes.[76] Due to Jamaica's geographical position,[77] it would appear that almost all of its provisions were imported from North America.

The North American colonies' increased importance in terms of victualling in the Caribbean theatre was extraordinary. What were the reasons behind this striking development? The first factor to be taken into consideration was the significantly shorter sea routes, thus allowing for better provisioning. Secondly, North America was financially more competitive, since its produce was cheaper than produce grown in England, and transport costs were lower.

70 PRO, ADM 110/18, Victualling Board to the Navy Treasurer.

71 PRO, ADM 110/18, 13 January 1757; ADM 110/19, 28 October 1757.

72 PRO, ADM 110/20, 31 March 1760.

73 PRO, ADM 110/15, 11 September 1747.

74 In the spring of 1757, Admiral Frankland requested that J. Biggin deliver his produce to the island of St Christopher. Biggin refused. PRO, ADM 110/19, 3 June 1757.

75 PRO, ADM 110/18, 23 March 1757; ADM 110/20, 10 March 1762; ADM 111/45, 18 March 1757; ADM 111/53, 10 March 1762.

76 PRO, ADM 110/18, 15 April 1757; ADM 110/20, 10 March 1762; ADM 110/53, 25 June 1762.

77 Cf. Chapter 6.

Thirdly, merchants were almost always guaranteed greater security for their provisions when these came from the northern colonies.

Indeed, the great concentration of French ships in European waters necessitated that merchant ships leaving England be escorted from the start of their journey. Thus, apart from the fairly high risk of capture, there was also a certain loss of independence for contractors who had to time their shipments to coincide with the availability of an escort. Gradually, they began to be of the opinion that obtaining regular provisions from Europe for fleets stationed in the West Indies was too hazardous.[78] In contrast, the very small French naval presence in Caribbean waters was not enough of a threat to warrant an escort for the merchant ships from North America. In fact, any danger at all came from the privateers, against whom it was easier to defend oneself. This greater flexibility provided the contractors with the possibility to send provisions on a much more regular basis. Consequently, the fleet in Jamaica received its provisions on a monthly basis, indispensably necessary to ensure the quality of food that had a tendency to spoil very quickly in that climate.[79]

Despite the fact that the naval forces sent to the West Indies were victualled for the most part by private contractors, the Victualling Board did send some convoys of provisions. This was due to the fact that the private contractors were under contract to provide for all the sailors of all Royal Navy ships, regardless of the number, but they were not under contract to provide for any soldiers that the ships might also be transporting.[80] On the three occasions when large numbers of troops were transported with a view to making conquests, the Victualling Board was obliged to ensure their victualling, following the same procedure that was used during the previous war, namely, by waiting for the Lords of the Admiralty to announce their decision, chartering merchant ships at Gravesend, transporting provisions to Spithead, and then placing them under escort from Spithead to the West Indies.[81]

Two shipments were organised in order to victual the 7,000 soldiers who, under the command of Commodore Moore and General Hopson, captured Guadeloupe in 1759. The first shipment, carrying four months' worth of provisions, left Spithead on 10 February 1759 escorted by the *Hampshire*;[82] the second shipment, carrying six months' worth of provisions for 7,000 men, left at the beginning of May.[83]

78 PRO, ADM 110/19, 15 August 1757.

79 PRO, ADM 110/19, 15 August and 2 November 1757, 22 March 1762.

80 PRO, ADM 110/19, 17 May 1758.

81 PRO, ADM 110/19, 23 and 30 April 1759.

82 PRO, ADM 110/19, 8 and 16 March 1759.

83 PRO, ADM 110/19, 16 March and 30 April 1759.

The sheer quantities involved in this last shipment, plus the fact that two other shipments were being prepared simultaneously in order to victual the 20,000 men involved in operations in North America,[84] resulted in such a demand for bread that the Victualling Board bakeries were simply unable to meet the demand. The Commissioners thus turned to private suppliers, resulting in a labour dispute in which the Victualling Board bakers threatened to go on strike if their pay did not match that of the private bakers.[85]

Also, apart from the provisions that Admiral Rodney had taken on board upon leaving England in 1761, six months' worth of provisions were sent to the 10,000 soldiers who, under his and General Monckton's command, were preparing to attack Martinique.[86] The army of 11,000 men who captured Havana under the command of General Lord Albemarle were victualled by an imposing convoy of provisions that Admiral Pocock had transported from England, consisting of seven months' worth of rations for 16,000 men.[87] In the event that the troops should suffer from a lack of food, it was agreed that they would receive provisions from the Navy.[88]

The results

The imposing naval presence in the West Indies was thus reasonably well victualled, apart from two periods in which some serious difficulties were encountered. One period – in 1757 – involved problems at the level of victualling. The other period involved a problem in terms of human relations, involving Admiral Holmes and the agents of the contractor in Jamaica. This crisis is interesting because it would result in new obligations for the merchant in late 1761 to early 1762.

Before passing judgement on the efficiency of the private victualling system for vessels stationed in the West Indies throughout the war, it is important to examine both the reasons for and the consequences of the difficulties that were raised.

In the first half of 1757, first Townsend and then Cotes – who replaced Townsend in command of the naval forces in Jamaica – repeatedly complained to the Admiralty about shortages in provisions, particularly of bread.[89] These

84 PRO, ADM 110/19, 1 and 23 April 1759.

85 PRO, ADM 110/19, 27 April 1759.

86 PRO, ADM 110/20, 23 October 1761.

87 D. Syrett, *The Siege and Capture of Havana 1762* (London, Navy Records Society, 1970), vol. 114, no. 50. Admiralty to Pocock, 18 February 1762.

88 Syrett, *Siege and Capture of Havana*, no. 47. Secret instructions given to General Albemarle, 15 February 1762.

89 PRO, ADM 1/234, Townsend to the Admiralty, 19 February and 22 March 1757; ADM 1/235, Cotes to the Admiralty, 24 May and 7 October 1757, 9 December 1759.

shortages were the result of an embargo imposed by the governors of North America in response to the number of operations carried out in this zone, and of the high risk the ships ran of being captured. Auguste Boyd's agents did their best to explain the dramatic consequences of the embargo on the crews to the governors of New York and of Pennsylvania, to no avail. No merchant ship was allowed to set sail before the beginning of May.[90]

Deprived of victuals from North America since December 1756, the 5,000 sailors were nonetheless provided with locally bought provisions, even though the daily ration was substantially modified by the presence of hastily made bread prepared from flour obtained from yams.

In order to avoid a similar problem in the future, Rear Admiral Cotes recommended that from now on, half of all bread requirements should come from England.[91] The Victualling Board duly transmitted this recommendation to the supplier, who chose to ignore it. He maintained his belief that monthly round trips from North America were much more reliable than provisions sent from England, and contented himself with asking that orders be sent to the governors of the northern colonies stating that, in the case of a new embargo, ships carrying provisions should nonetheless be allowed to set sail.[92] The Commissioners replied favourably to Boyd but demanded, on the other hand, that he always have four months' worth of stocks of bread in his warehouse in Jamaica.[93]

Despite the raising of the embargo there was nonetheless an extreme shortage of peas in Jamaica towards the end of 1757. This was partly due to the fact that a ship carrying provisions had been captured, and partly because the departure date for another convoy from North America was delayed due to a lack of escort ships, resulting from a more pronounced French presence in those waters.[94] However, this dearth of peas did not have a detrimental effect on the men's victualling since Rear Admiral Cotes ordered that the peas could be replaced by rice. This decision did, however, result in friction between Cotes and the Victualling Board. Since rice was more expensive than peas, Mr Boyd asked that he be reimbursed the difference of £212, a request that was refused by the Commissioners who were of the opinion that the Royal Treasury should not have to suffer as a result of the negligence of a contractor who failed to send a sufficient quantity of peas.[95]

90 PRO, ADM 110/18, 15 April 1757; ADM 110/19, 15 August 1757.

91 PRO, ADM 1/235, Cotes to the Admiralty, 24 May 1757.

92 PRO, ADM 110/19, 15 August 1757.

93 PRO, ADM 110/19, 2 November 1757.

94 PRO, ADM 110/19, 6 July 1757.

95 PRO, ADM 1/307, Douglas to the Admiralty, 19 August 1760; ADM 110/20, 15 December 1760.

The second period of tension was due to totally different reasons. It all began in September 1761 when Messrs Meylor, Halls and Torry – agents to the contractor in Jamaica – let it be known to Admiral Holmes that from now on they were refusing to supply him with weekly bulletins stating which victuals had been supplied to the ships and what still remained in stock. Holmes had requested this information when he was first posted to Jamaica in May 1760 and had received it regularly since then. He wrote to the Commissioners requesting that the agents be discharged and asking that from now on a clause requiring strict obedience on the part of a contractor's agents towards a naval commander in chief be included in the contract.[96]

The role played by the Victualling Board in this crisis underlines the Commissioners' desire to ensure that all contracts drawn up between them and a contractor were strictly respected, whilst at the same time allowing the contractors as much flexibility as possible. In this case, the Commissioners sided with the contractor and advised the Lords of the Admiralty against introducing the clauses requested by Holmes:

> Upon the receipt of which letter we sent to Messrs. Nesbitt and Franks, the contractors, and discoursed with them upon the said affair, and they acquainted us a letter which they sent us upon the said occasion (and of which we herewith enclose a copy) that the discharging of their agents was a step that the present situation of affairs in Jamaica would not admit of but nevertheless they had given directions to them to make a proper submission to the Admiral for any slight and to deliver the accounts in future although their contract did not oblige them to it. We are humbly of the opinion that it will not be advisable to insert any clause in a new contract dismissing their agents it not being to expect that any persons will enter into a contract under the condition of trusting their property and the management of their affairs abroad in the hands of any agents but what are of their own choosing.[97]

In their reply to the contractor, the Lords of the Admiralty stated that they hoped that in the future the contractor would oblige his agents to provide whatever information a squadron commander might ask for.[98]

Clearly annoyed by a letter sent by Captain Forrest – Holmes's successor in Jamaica following the latter's death – in which he referred to a bread shortage in December 1761, the Commissioners succeeded this time in imposing a new clause when the time came to renegotiate their contract with the same contractors, namely Nesbitt and Franks. The new clause was severe, but undoubtedly advantageous for the best possible victualling of the Navy. Its

96 PRO, ADM 110/20, 18 January 1762.

97 PRO, ADM 110/20.

98 PRO, ADM 110/20, 22 March 1762.

essence is conveyed in the letter reprinted below from the Commissioners to the Lords of the Admiralty:

> Their agents are obliged to deliver the commanding officer at Jamaica for the time being an account weekly or whenever he shall require the sum of provisions remaining in their stores there by means of which he will be able to see from time to time whether they are sufficiently stocked to answer the demand of His Majesty's ships and we have wrote to Captain Forrest to desire in case he should find them short and the contractors agents should refuse or neglect to take timely and effectual measures to procure the supplys which he may judge to be absolutely necessary for carrying on His Majesty's Service that he will purchase in the best manner and on the best terms he can what provisions he may find wanting for that purpose and draw bills upon us for the amount thereof and if the Crown should be put to any extraordinary expense thereby we shall abate the same from the contractors bills or take such other methods as may be most proper to indemnify His Majesty.[99]

Suppliers in London were also subject to strict surveillance by the Commissioners, who regularly enquired about what measures the suppliers were taking to ensure optimal victualling of the naval forces in the West Indies. This could involve asking for a detailed list of provisions already sent to the West Indies, and of those waiting to be sent.[100]

In conclusion, it would appear that the competitive, pragmatic and empirical system that was established in the yards in the West Indies under the jurisdiction of the Victualling Board in the course of the period examined, and especially during the Seven Years War, could be considered to be a reflection of North America's economic development in the same period. Two figures clearly illustrate this evolution: in the 1730s, the cost of victualling *per capita* in the West Indies was eleven and a quarter pence;[101] in 1760, during the climax of the war and in the same geographical zone and with a record number of men involved, the cost of victualling *per capita* had dropped to nine pence.[102]

The results which have surfaced in this study become meaningful in the light of the evolution of the victualling of French naval forces sent to the West Indies in the same period (this data has been studied and commented on in the author's doctoral thesis).[103] Progress made in the British logistical system meant that the transport of victuals became more and more efficient,

99 PRO, ADM 110/20, 22 March 1762.

100 PRO, ADM 111/54, 5, 9, 19, 21 and 23 July, 15 September, 18 October, 8 November 1762.

101 Ranft, *The Vernon Papers*, 358–60. Mason and Simpson to the Victualling Board, 20 November 1739.

102 PRO, ADM 111/50, 7 March 1760.

103 Buchet, *La lutte*, II, 924–1092.

permitting better victualling of growing numbers of men. In contrast, the French contractors seemed unable to abide by the stipulations laid out in their contracts concerning the victualling of French forces in the West Indies. This inevitably led to a decrease in the amount of time a fleet could spend in Caribbean waters, and a limit on the number of men that could adequately be provided for. These shortcomings were undoubtedly at the heart of the defeats suffered by the French in this theatre during the Seven Years War.

PART THREE

THE MAIN MARKETS

8

Meat

Victualling the Royal Navy with meat was a very delicate operation, both due to technical problems associated with preserving it (see Chapter 3), and because significant quantities were required. Compared to the already large quantity of beef bought in England during the War of the Austrian Succession in 1746, namely 69,357 hundredweight (one hundredweight (cwt) = 112 lbs), the Commissioners had to find almost twice this quantity in 1759, namely 123,552 hundredweight. The increase is similar when the comparison is extended to the last four years of the two wars: 235,941 cwt of beef consumed during the War of the Austrian Succession as against 432,061 cwt of beef consumed during the Seven Years War.[1]

These figures, which only relate to English-bought beef, and do not include imports of Irish beef which also vastly increased during the Seven Years War, clearly show the unprecedented extent to which the Victualling Board's task grew in this period of hostilities.

Since the efficiency with which the Board operated cannot just rest on figures concerning quantities, the first part of this chapter will investigate fluctuations in price, and this will be followed by an examination of the extent to which the market was fragmented, as illustrated by a study of the body of contractors.

Everyday Functioning and Costs: The Difficulties

There existed two main types of contract: one for supplies of fresh meat, and one for supplies of salt meat.

Fresh meat contracts were drawn up for quarterly periods, the price per hundredweight having been agreed upon at auction, and the contractor was under the obligation to provide whatever quantities might be asked of him, within seventy-two hours.[2] The contracts covered five geographical zones, corresponding to the principal ports, and adjacent anchorages where

1 PRO, ADM 110/22, 18 February 1765.

2 PRO, ADM 110/46, 23 May 1757.

a cruising ship could easily drop anchor for as long as it took to take provisions on board, without needing to enter the port. These five zones and their base ports were as follows: the Thames (London); Chatham–The Nore–Sheerness–Blackstakes–Seareach; the Downs (Dover); Portsmouth; and lastly Torbay as of May 1759,[3] the excellent anchorage near Plymouth.

The contracts were always drawn up on the last day of March, June, September or December and came into effect on the first day of the following month. The only exception was for contracts drawn up for Portsmouth, which ran a month in advance, so that contracts drawn up on the last day of February, May, August or November came into effect on the 1st of March, June, September or December.

There was also one difference in terms of the content of the contracts: those drawn up for the Thames, Dover and Portsmouth specified beef, whilst those drawn up for Torbay specified beef and mutton and those for Chatham–The Nore–Sheerness–Blackstakes–Seareach specified mutton only. Furthermore, sheep destined for Chatham had to be slaughtered in Chatham, whereas sheep destined for The Nore–Sheerness–Blackstakes could be slaughtered only at Leigh or at Sheerness, and sheep destined for the ships at Seareach (that is to say in the offing of the bay formed by these four places) could be slaughtered '*where it may be most convenient*'. This last concession did not represent any form of financial loss for the Crown since, in each of the five zones and for all types of meat, the contractors were bound to deliver the meat directly on board the vessels.

A sixth and final type of contract was introduced at the end of January 1757. This was also a quarterly contract and was specifically concerned with the provision of fresh mutton (called 'wether mutton'[4]) for the Sick and Wounded Board, and used for the preparation of portable soup.

Table 8.1 is based on information taken from a report written in November 1761 by the Victualling Board and addressed to the House of Commons. Even though it only covers deliveries of fresh beef, it gives an insight into the very large quantities that were being dealt with.

The four places mentioned in the table thus received a total of 53,453 hundredweight of fresh beef in 1759, a figure rising to 70,824 cwt the following year. This represented the slaughter of 9,133 head of cattle in Portsmouth, Plymouth and Dover in 1759 and 12,259 head of cattle in 1760. Even though Portsmouth received the largest quantities of fresh meat in 1759 and 1760, Plymouth's growth in the same period was proportionately greater – to the

3 The first contract found was dated 21 May 1759: PRO, ADM 111/49.

4 A wether is a castrated ram. I am grateful to Dr Janet Macdonald for bringing this to my attention (Macdonald, 'A new myth of naval history?', *International Journal of Maritime History*, XXI (2009), 178, n.75).

Table 8.1 Quantities of fresh beef delivered to the principal yards

	1759			1760			1761[1]	
Yard	No. head of cattle slaughtered	cwt	%	No. head of cattle slaughtered	cwt	%	No. head of cattle slaughtered	cwt
Thames (London)	?	760	1.42	?	1,761	2.49	?	1,025
Portsmouth	4,963	27,268	51.01	6,046	32,843	46.37	?	23,709
Plymouth-Torbay	3,454	21,133	39.54	5,351	31,047	43.84		24,213
Downs (Dover)	716	4,292	8.03	862	5,173	7.30	?	?
		53,453	100		70,824	100		

1 Until 21 November.

Source: Table drawn up according to data in ADM 110/20, ff. 396–9.

point that in 1761 it overtook Portsmouth in terms of victualling fresh meat, mainly because it had the responsibility for victualling the fleet and Belle-Ile expedition off the French coast. The much lower figures as far as London is concerned clearly illustrate its different needs as one of the Royal Navy's primary victualling centres, but with only a secondary role in the fitting out and decommissioning of warships. As it happens, it was much more efficient and profitable not to bring the cattle to London, but instead to make good use of regional possibilities.

Figure 8.1 superimposes price fluctuations on a geographical zone-by-zone basis and reveals a number of interesting financial points. It is necessary, however, to ignore the bigger price differences which are due to seasonal variation. In his study of prices in England, W. Beveridge noted a sharp increase in the spring as opposed to the autumn, which was the season during which cattle and sheep could be bought for the best prices.[5] Figure 8.2, however, which is based on annual averages, shows a downward trend. This is because prices rose until the winter of 1758–9 and then began to drop somewhat, in spite of the increasing numbers of men that the Victualling Board had to provide for in the last years of the war. From a comparative point of view, the cheapest station at which to buy beef was London (Thames), followed by the Downs (Dover) and then Portsmouth (see Figure 8.2).

The provision of beef and pork for salting purposes followed a process that had been progressively established during the forty years preceding the Seven Years War. The poor quality of salt meat during the War of the Spanish Succession has already been discussed in the author's doctoral thesis: much of

5 Beveridge, *Prices and Wages*, 551.

Figure 8.1 Fluctuations in the prices of fresh meat (quarterly movement)

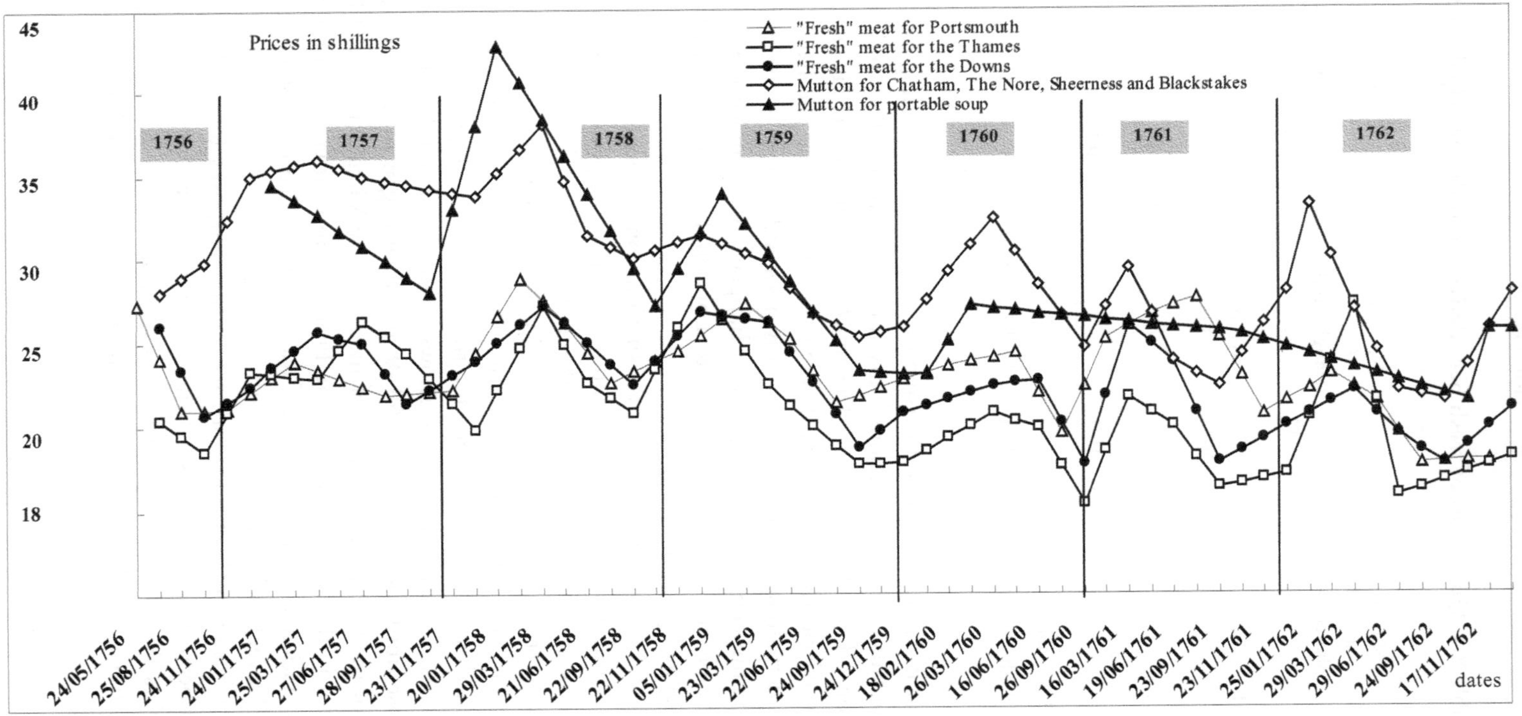

Figure 8.2 Annual average fluctuations in the prices of fresh meat

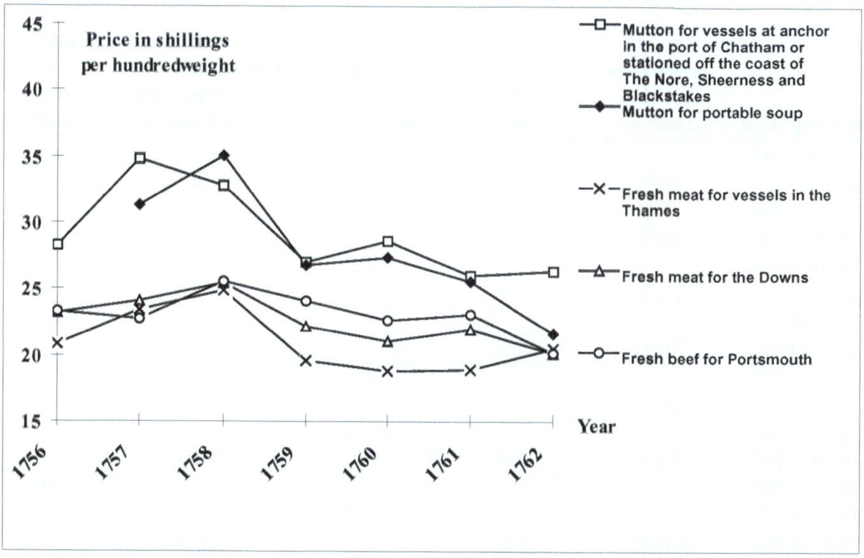

the meat was found to be rotten just a few months after having been cured. The well-justified recriminations on the part of the Admirals resulted in a number of investigations which revealed that most of the poorly preserved meat had been cured in the summer months. Consequently, the Victualling Board issued strict directives regarding the months in which meat could be salted. As of the 1739–48 war, meat could only be cured from the end of September until the beginning of May, '*as ye weather ... permitted and ye service required it*'.[6]

This salting timetable was even more strictly adhered to during the Seven Years War and gave rise to two quarterly contracts for the ports of London, Portsmouth, Plymouth and Dover, concerning the dates by which the livestock had to be delivered: the first contract was applicable from the last days of September or at the latest the beginning of October, the second contract about the 20th of December, and the livestock had to be delivered by the 31st of December and the 31st of March respectively.

Merchants in competition for contracts in Portsmouth, Plymouth and Dover were asked to declare the number of pigs and head of cattle they felt confident they could deliver, plus their asking price per hundredweight. Those hoping to win a London contract had to ensure that they could provide a fixed number of livestock on a twice-weekly basis. On a pre-determined day,

6 Baugh, *British Naval Administration*, 408.

Table 8.2 The prices of fresh meat bought by the Commissioners of the Victualling Board (in shillings per cwt)

Bidding date	Fresh beef for the Thames	Fresh beef for the Downs	Fresh beef for Portsmouth	Mutton for Chatham, Nore, Sheerness and Blackstakes	Mutton for portable soup
24/05/1756			27/4		
25/06/1756	20/5	26		28	
25/08/1756			21		
27/09/1756	18/7	20/9		28	
24/11/1756			21/2		
29/12/1756	23/4	22/6		28	
24/01/1757					34/6
25/03/1757	22/11	25/8		36	
23/05/1757			?		
27/06/1757	26/4	25		28	
24/08/1757			21/11		
28/09/1757	24/5	21/5		34/5	
19/10/1757					28
23/11/1757			22/3		
21/12/1757	19/10	23/11		33/9	
20/01/1758					42/10
22/02/1758			28/10		
29/03/1758	27/2	27/2		38	
24/05/1758			26/2		
21/06/1758	22/7	25		31/5	
23/08/1758			22/8		
22/09/1758	20/10	22/6		30	
27/09/1758					27/2
27/12/1758	28/6	26/10		31/5	
05/01/1759					27/2
21/02/1759			27/4		
23/03/1759	22/6	26/2		31/5	
16/05/1759			25/2		
22/06/1759	20	22/8		26/8	
17/08/1759			21/5		
24/09/1759	17/9	18/9		26/8	23/4
16/11/1759			22/4		
24/12/1759	17/10	20/10		25/11	
31/12/1759					23/1
18/02/1760			23/8		
03/03/1760					27/2
26/03/1760	20/10	22/5		32/5	
21/05/1760			24/5		
16/06/1760	19/11	22/9		28/5	
20/08/1760			19/8		
26/09/1760	15/5	17/9		24/9	
18/02/1761			25/3		
16/03/1761	21/9	26		29/5	

20/05/1761			26/9		
19/06/1761	20	23/10		23/11	
24/06/1761			19/3		
23/09/1761	16/4	17/10		22/5	
07/10/1761					25/6
23/11/1761			20/9		
23/12/1761	17/2	20		28	
25/01/1762				33/2	
22/02/1762			23/1		
20/03/1762	27/2	22/1		26/11	
24/05/1762			21/7		
23/06/1762	15/10	19/4		22/1	
20/08/1762			17/9		
24/09/1762		21		27/11	
13/10/1762					21/7
17/11/1762			18		
24/12/1762	18/2	21		27/11	

the Commissioners would summon all the merchants to appear together and would announce the number of pigs and head of cattle to be delivered every week. For example, on 27 September 1758 the Commissioners announced to the merchants that their needs would require the slaughter of 120 head of cattle and 400 pigs on a twice-weekly basis from 1 October to 31 December, thus adding up to a total of 2,980 head of cattle and 3,600 pigs.[7] All merchants who considered that they could fulfil the demand were individually asked to disclose their prices in writing, having been informed that the most financially competitive proposal would be chosen. Unlike the process for other victuals, the merchants would not have recourse to a second summons.

To ensure the smooth running of the slaughterhouses, pigs had to delivered on Tuesdays and Thursdays; cattle on Mondays and Fridays. Cattle could not weigh less than 7 cwt and pigs no less than 1 cwt. If all the prices were judged to be too high, then the Commissioners could modify the selection procedure. For example, in October 1757, when faced with exorbitant pork prices, the Commissioners issued contracts of one month's duration only. In this way, there was a monthly renewal of contracts throughout the long winter of 1757–8.[8]

A report addressed to Parliament in February 1765 reveals how many hundredweight of beef were bought by each of the four main ports between 1 January 1745 and 31 December 1764. Figure 8.3 is based on this information and reveals a number of interesting points. First of all, there is a

7 PRO, ADM 111/49, 27 September 1758.

8 PRO, ADM 111/46, 12 and 26 October, 23 November, 23 December 1757; 9, 25 and 30 January 1758.

Figure 8.3 Beef bought on an annual basis for the ports of London, Portsmouth, Plymouth and Dover between 1 January 1745 and 31 December 1764

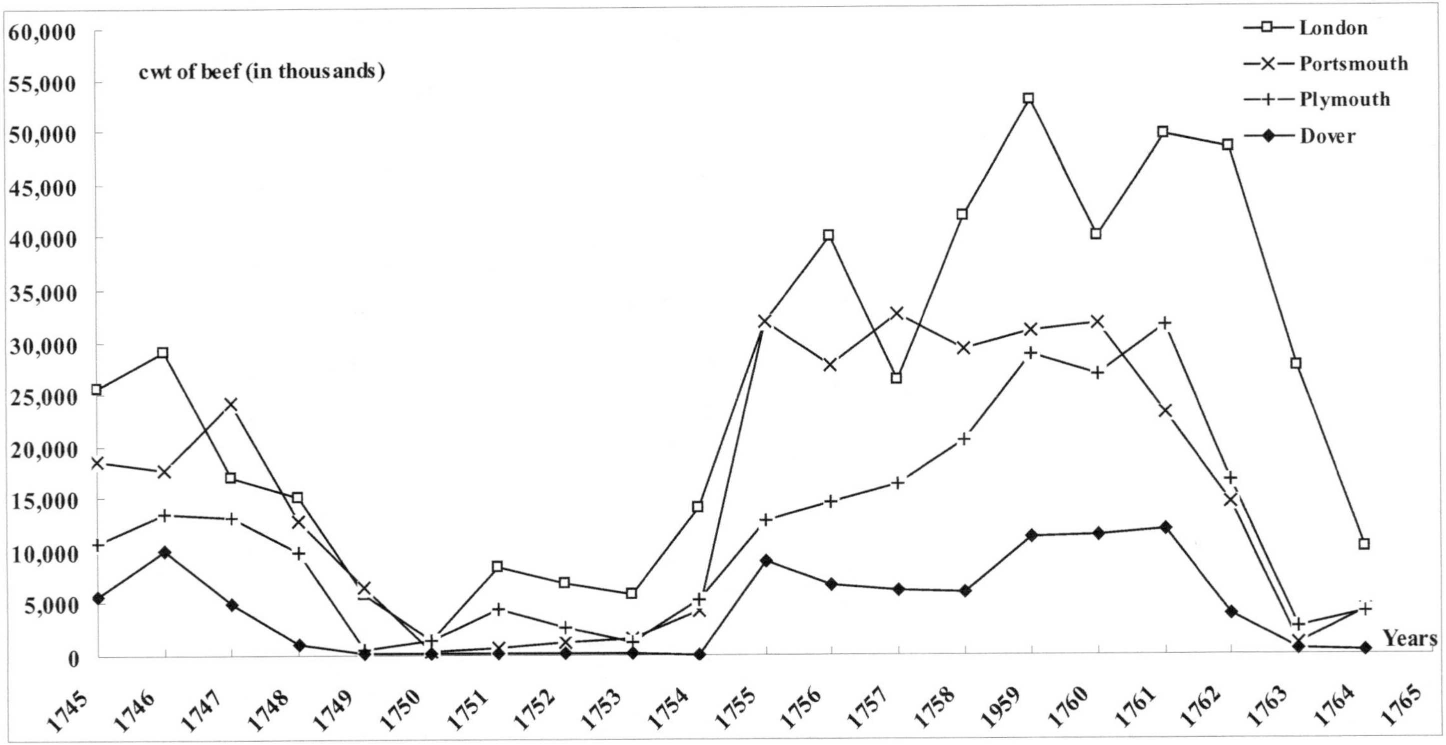

Source: PRO, ADM 110/22, Report from the Victualling Board to Parliament, 18 February 1765.

Figure 8.4 Royal Navy manpower: numbers mustered

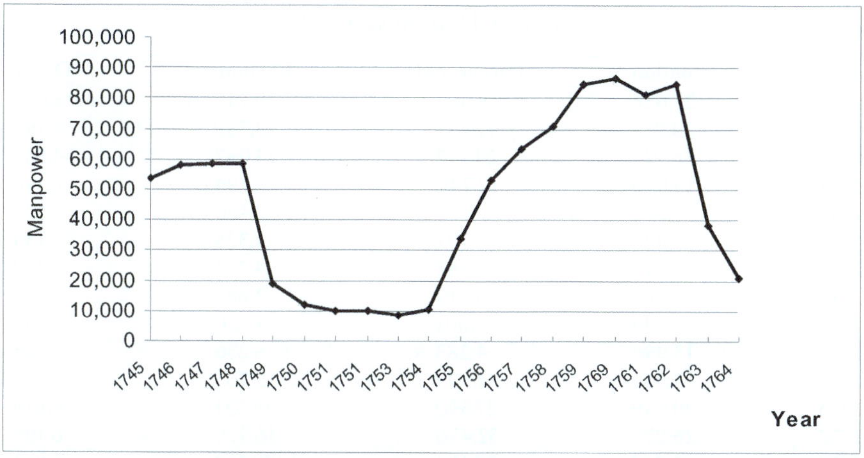

Source: N.A.M. Rodger, *Command of the Ocean, 1649–1815* (London: Allen Lane-Penguin, 2004), 637–8.

direct correlation between the curve of this graph and that constructed from naval manpower numbers given in N.A.M. Rodger's *Command of the Ocean* (Figure 8.4). Figure 8.3 clearly illustrates the extent to which the Royal Navy anticipated the coming manpower increases in terms of preparation for the war. Even though the Seven Years War was not declared until 1756, by 1754–5 Navy logistics personnel were already getting ready for the war since the mobilisation of seamen necessitated, as a preliminary measure, that the warehouses be filled with provisions. Secondly, the rise seen for Plymouth in Figure 8.3 and as detailed in Table 8.2 illustrates the gain in importance of this port which, in 1761, overtook Portsmouth in terms of the total volume of provisions bought. Finally, 1759 and 1761 were clearly years in which the largest quantities of beef were purchased. This observation will lead us to the next section, namely the analysis of price increase.

Figure 8.5 is based on averages calculated in the winter for the prices of meat destined for salting. The graph reveals that the tensest moment in terms of pork prices occurred during the winter of 1757–8. It will be remembered that it was necessary to have recourse to monthly contracts drawn up in October, November and December 1757 (see Table 8.4 which lists successive prices for pork purchased in the different ports). In January 1758, the Commissioners had no choice but to face the facts: the price of pork, which had now reached 45s 3d per cwt, represented an increase of over 30 per cent on the highest price of the preceding year. It was now too expensive to

Table 8.3 Cwt of beef bought on an annual basis for the ports of London, Portsmouth, Plymouth and Dover between 1 January 1745 and 31 December 1764

Year	London	Portsmouth	Plymouth	Dover
1745	25,626	18,569	10,747	5,602
1746	29,052	17,637	13,389	4,886
1747	16,914	24,172	13,087	4,853
1748	15,026	12,746	9,794	1,096
1749	5,696	/	460	120
1750	1,442	307	1,324	130
1751	8,440	662	4,354	119
1752	6,767	1,119	2.685	130
1753	5,734	1,601	1,160	90
1754	13,996	4,283	5,236	26
1755	31,923	31,909	12,705	8,910
1756	39,949	27,590	14,521	6,619
1757	26,277	32,456	16,279	6,199
1758	41,882	29,184	20,512	5,861
1759	52,742	30,965	28,650	11,195
1760	39,847	31,639	26,691	11,327
1761	49,447	23,031	31,570	11,903
1762	48,232	14,469	16,543	3,811
1763	/	985	2,570	450
1764	10,059	4,266	4,035	379
Total	469,859	307,775	236,320	84,730

Source: PRO, ADM 110/22, Report from the Victualling Board to Parliament, 18 February 1765.

Table 8.4 The price, in shillings per cwt, of pigs bought in the ports of London, Portsmouth, Plymouth and Dover whose meat was destined to be salted

	London	Portsmouth	Plymouth	Dover
Winter 1756–7	33s 4d – 35s 4d	34s – 34s 11d	29s 8d	29s 10d – 32s 10d
Winter 1757–8	34s 9d – 34s 3d – 36s 3d 45s 3d (Prices too high – no contract)	35s	/	41s (Prices too high – no contract)
Winter 1758–9	43s 6d – 43s – 43s 9d	42s 5d	/	/
Winter 1759–60	36s (Prices too high – no contract) 33s 6d – 30s 5d – 27s 11d 35s 3d (Prices too high – no contract)	34s	30s 8d	32s
Winter 1760–1	?	?	?	?
Winter 1761–2	29s 4d – 31s 8d – 35s 6d	30s	23s 6d – 23s 6d	23s 11d
Winter 1762–3	29s	/	/	/

Source: Data taken from the ADM 111 series.

Figure 8.5 Fluctuations in the prices of meat bought for salting purposes

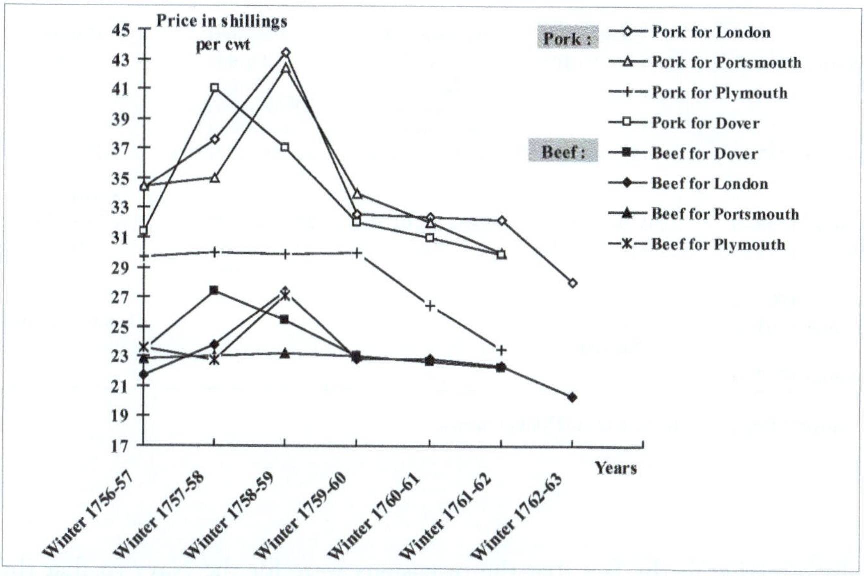

continue buying it. In Dover, at the same period and for the same reason, it was decided to stop buying pork.

The following winter of 1758–9 was also very difficult and saw inflated prices, even though the outrageous figure of 45s 3d per cwt was not attained again (see Table 8.3). On the other hand, prices did drop substantially over the next few years, until they levelled out at slightly below those in force up until then.

A similar pattern emerged (see Table 8.5) concerning the price of beef. These prices also climbed rapidly during the winter of 1758–9, but unlike the situation for pork, beef prices peaked during the winter and then followed the pattern for pork and dropped again during the last four years of the war (see Figure 8.5).

These reductions in prices were all the more notable since the number of men to victual continued to rise until 1762. There is a fairly simple reason for the reductions: the Commissioners of the Victualling Board succeeded in controlling prices because they were finally able to count on meat imports from Ireland – a novel situation, because it had been illegal to import any Irish agricultural product into England since the reign of Charles II. Until then, only far-distant English possessions such as the West Indies and North America had been allowed provisions of meat, butter and cheese from Ireland. The contractors did not fail to resort to this since they could take advantage of the lower prices. Nonetheless, they required a certificate from the Admi-

Table 8.5 The price per cwt of cattle bought in the ports of London, Portsmouth, Plymouth and Dover whose meat was destined to be salted

	London	Portsmouth	Plymouth	Dover
Winter 1756–7	21s 8d – 21s 10d	22s 9d 22s 11d (Price too high – no contract)	23s 9d 23s 6d (Price too high – no contract)	23s 5d
Winter 1757–8	23s 9d	21s 4d – 24s 8d	22s 8d – 22s 10d	27s 5d (Price too high – no contract)
Winter 1758–9	24s 9d – 30s	19s 10d – 26s 5½d	26s 11d – 27s 3d	23s 7d – 27s 3d
Winter 1759–60	22s 9d – 21s 5d – 24s 4d	23s	(Price too high – no contract)	23s
Winter 1760–1	?	?	?	?
Winter 1761–2	22s 4d – 21ss – 23s 10d	/	/	21s 8d – 22s 10d
Winter 1762–3	20s 4d	/	/	/

Source: Data taken from the ADM 111 series.

ralty attesting to the fact that the provisions were for the Navy so that the embargo could be broken. In fact, even before war was declared, in September 1755, the Lords of the Admiralty prepared for all contingencies by sending a Victualling Board Commissioner, Horatio Townshend (see Appendix 2), to Ireland so that he could verify that Irish salt meat corresponded to the quality norms required by the Royal Navy.[9] As of 1756, the Navy had bought and dispatched 6,400 barrels of beef and 3,400 barrels of pork from Ireland to Gibraltar.[10] However, it was undoubtedly the exponential increase in the price of pork on the London market during the winter of 1757–8, resulting in the Commissioners abandoning their purchases of pork in February 1758, which was the true reason for the radical changes that occurred. The following month, in March, the Lords of the Admiralty succeeded in persuading Parliament to relax their protectionist measures. Imports of salted meat and butter from Ireland were now permitted for a period of six months, taking effect from 24 June. Without waiting for 24 June, the Victualling Board drew up a contract with James Benson on 14 April for the victualling of ships stationed in Spithead, Portsmouth and Plymouth with 2,000 barrels of pork. These provisions were to be loaded directly onto the ships, without passing via the yard warehouses, in an effort not to violate a law that was still in force.

At the end of the six-month period, in December 1758, it was decided that the new legislation should be renewed on an annual basis until December

9 PRO, ADM 111/46, 16 July 1756.

10 Calculations based on the ADM 111 archival series.

Table 8.6 Salt beef imported from Ireland

	Winter 1756–7	Winter 1757–8	Winter 1758–9	Winter 1759–1960	Winter 1760–1*	Winter 1761–2
Pork	/	2,000 barrels	8,000 barrels	6,000 barrels 1,170 puncheons 2,600 tierces[1] (842,400 pounds) 100 hoghead[2] (44,000 pounds)	3,000 tierces (972,000 pounds) 4,000 barrels	10,500 barrels 600 tierces[1] (194,400 pounds)
Beef	/	/	/	3,200 barrrels 174 hogheads	4,180 tierces[3] (1,387,760 pounds) 3,600 barrels 114 puncheons[4] (71,820 pounds)	1,000 tierces[3] (332,000 pounds)

* (Archival gaps → incomplete data)

1 1 *tierce* of pork = 324 pounds of meat (1 tierce = 76 pieces)
2 1 *hoghead* of pork = 440 pounds of meat
3 1 *tierce* of beef = 332 pounds of meat
4 1 *puncheon* of beef = 630 pounds of meat

Source: Data taken from the ADM 111 series.

Table 8.7 Average prices in the years 1745–64 for fresh meat bought in the main ports

Port	Beef *per cwt*	Pork *per cwt*
London	21s/10 229/448	30s/1 132/448
Portsmouth	21s/11 24/448	30s/2 342/448
Plymouth	20s/7 342/448	26s/3 239/448
Chatham	22s/10 233/448	/
Dover	22s/4 120/448	33s/6 340/448

Source: ADM 110/ 22, Report written for Parliament, dated 12 February 1765.

1761. After this date, the Victualling Board alone was authorised to import food provisions from Ireland.

Consequently, from 15 November 1758 until the end of the war, the Commissioners of the Victualling Board were able to supplement their meat supplies from Ireland with the quantities they considered to be necessary. In 1758 this only applied to pork, but from the autumn of 1759 onwards, the sharp increase in the price of English beef meant that Irish salt beef was transported in ever-increasing quantities to English yards (see Table 8.6). By using the Irish and English markets, the Commissioners found themselves in the position of being able to diversify their sources and use the competitiveness of the two parallel markets to their advantage.

In a statement given by the Victualling Board to the Lords of the Admiralty on 12 February 1765 was a list showing the average price of fresh

English pork or beef bought in the main ports over the past twenty years – namely, from 1745 to 1764. That the list was requested reveals the importance accorded by the Lords to logistical issues, and of even greater interest here is that it reveals price fluctuations between ports (see Table 8.6).

The table reveals that beef prices remained generally consistent throughout this period, apart from an almost insignificant rise in price in the Chatham and Dover ports and a very slight drop in price at Plymouth. Pork, however, revealed bigger differences in rates. Average prices at Dover and Plymouth were diametrically opposed, and quite different from the going rates in London and Portsmouth. It is very likely that the more competitive prices for meat in Plymouth were a factor in this port's remarkable development.[11]

The Contractors

The body of contractors who participated in the Victualling Board's call for tenders for the victualling of the Royal Navy in meat provisions consisted of 122 individuals and/or associations. This figure can be broken down as follows: for English meat, 94 individual contractors and 28 associations (in which there appeared 13 new names), making a total of 107 people, to whom must be added 14 candidates for the supply of Irish meat.

In terms of English meat only, there were 195 contracts; 92 per cent of these were won by individual contractors (180 markets), whilst only 15 contracts were won by 28 associations of merchants. However, it would be erroneous to minimise their economic importance, for the following reasons: first, nine of the fifteen contracts were for salt meat only; second, these associations and the complicity between merchants that they implied have permitted the discovery of certain commercial networks which question the degree to which the market was really broken up.

Before an analysis of the contracts issued for salt meat and fresh meat can begin, a couple of observations need to be made on the ninety-four individual contractors.

First of all, regional specialisation amongst these merchants was most pronounced. If the five major ports are taken into consideration – London, Portsmouth, Chatham, the Downs (including Dover), and Plymouth/Torbay – then it is noticeable that fully 78 per cent of the merchants (seventy-three individuals) participated in tenders concerning only one geographical zone

11 The development of livestock breeding in Devon and Cornwall explains the more competitive pricing of meat in this port. Since the 1750s, the southwest had succeeded in even satisfying London and the southeast's meat requirements. See Thirsk, *The Agrarian History*, vol. V.2, 358–82. See also vol. V.1, 872, for a table showing beef prices per region.

Table 8.8 A breakdown according to the region the merchant traded in

Merchants who participated in invitations to tender bids in one geographical area only

London	No. of contracts won	Portsmouth	No. of contracts won	Plymouth-Torbay	No. of contracts won
Tho. Ayres		Tho. Best		Edward Brown	
T. Bell		RobertFowler Coade	1	J. Chilcot	
D. Bisson Junior		Thomas Haine		William Clark	
John Cook	10	Robert Harding		William Colten	1
Edward Edes		Samuel Johson		Richard Cross	10
John and Peter Lefebvre		William Lea	4	John Culine	1
Isaac Lefebvre		James Perry		Peter Culune	
Lloyd		Thomas Purdery		Richard Disting	
Joseph Mawbey		Daniel Witean		William Earled	
Daniel Ponton		Robert Willian		Thomas Edmunds	
Edmond Russell				Thomas Edward	2
Henry Wilkes				Edward Fossett	1
Roger Wright				Francis Giles	
	Contracts		**Contracts**	John Jesseps	
Dover-Downs	**won**	**Chatham**[1]	**won**	Thomas Jewell	1
Sam. Brothers		Albin		William Lake	
Nich. Burvit		David Harridge	1	John Lear	
John Cannon		Henry King	7	Edward Lee	
Ralph Chambers	1	Richard Martin	1	John Littleton	
Martin Child	2	Harris Roughton		William Metherell	5
Thomas Henley		Thomas Scott		Richard Phillips	
Matthew Kennet	1	Edward Worth		John Snell	
Pheneas Kite				John Spurrel	
William Lord				Thomas Symons	1
Edward Monns				John Troude	1
Edward Mount				John Wills	
Henry Sayer	3			Richard Windsor	
Thomas Toil					
Nathanael Walker	6				
Thomas Webb					

Merchants participating in invitations to tender bids in two geographical areas

James Adams (London and Plymouth)	Joseph Fysson (Plymouth and Portsmouth)
John Pass (London and Plymouth)	Richard Joliffe (Plymouth and Portsmouth)
THurford (London and Portsmouth)	William Stevens (Plymouth and Portsmouth)
Georges Lewis (Dover, Downs and Chatham)	Richard Turner (Plymouth and Dover-Downs)
Edward Martin (Dover, Downs and Chatham)	

Merchants who operated in a diversified geographical field

James Berry (London, Chatham, Portsmouth, Dover-Downs) = All zones except Plymouth
Joseph Brownsworth (London, Chatham, Portsmouth, Dover-Downs) = All zones except Plymouth
Georges Carter (Chatham, Portsmouth, Dover-Downs)
Georges Deake (London, Chatham, Dover-Downs)
William Ford (London, Chatham, Portsmouth, Dover-Downs) = All zones except Plymouth
Thomas Lewis (London, Chatham, Portsmouth, Dover-Downs) = All zones except Plymouth
Peter Mellish: All zones
Samuel Mellish (London, Chatham, Portsmouth, Dover-Downs) = All zones except Plymouth
Johnson Pasco (London, Chatham, Portsmouth)
William Preddy (Chatham, Portsmouth, Dover-Downs)
Allen Spencer (London, Chatham, Portsmouth, Dover-Downs) = All zones except Plymouth
John Woodard (London, Chatham, Portsmouth, Dover-Downs) = All zones except Plymouth

Total number of individual merchants: 94

1 Contracts for fresh meat for the Thames have been included in the Chatham geographical zone.

(see Table 8.8). Only nine merchants tendered for two zones, and only 13 per cent of all the merchants (twelve out of a group of ninety-four) actually tendered over a large-scale geographical area.

Secondly, this type of victualling had a highly concentrated profile. In fact, thirty-six merchants would win the 180 available contracts, and of these thirty-six, the thirteen most important among them would win 148 contracts, representing 82 per cent of the share (see Table 8.8). Six of these merchants stood out from the others in terms of sheer importance. The first four, Thomas Lewis, Samuel Mellish, Allen Spencer and James Berry, had bid for all the zones under consideration except the Plymouth/Torbay zone, which was an outlying zone, geographically speaking. They succeeded in respectively acquiring 37, 21, 14 and 12 contracts. The remaining two merchants both had highly specialised profiles. Richard Cross was geographically specialised, and only bid for the Plymouth/Torbay zone, regardless of whether the contract was for salted or fresh meat. He won ten contracts. John Cook had an even more specialised profile, and only dealt in salt pork in the London zone. He bid for, and won, the ten available contracts (see Tables 8.8 and 8.9).

Research carried out on the main contractors in the Corporation of London Record Office archives revealed disappointingly little. There is only information of any substance on William Preddy who won contracts for Portsmouth, London, Chatham and Dover. The archives contained his certificate of admission issued in 1742, found in the register listing the names of people who had obtained the London City Freedom Admissions. He first worked as an apprentice for a master feltmaker, John Messenger, but most significantly, he was the son of Henry Preddy, master butcher in Bermondsey.[12] Consequently, he must have been fairly comfortable with regularly participating in the invitations to tender bids to the Commissioners of the Victualling Board.

The fact that no other contractors appeared in the City Freedom Admissions register may indicate that the other contractors responsible for the Navy's provisions of meat did not operate within the capital.

Suppliers of salted meat

The port of London market, which was the most important by far since it also supplied part of the victualling of other ports, at first glance appears to have attracted the highest number of bidders (twenty-two individuals and sixteen associations) and consequently to have benefited from widely spread contracts (Table 8.10: A). However, a study of the breakdown of contracts obtained, contractor by contractor, would suggest taking a more cautious view. In fact

12 Corporation of London Records Office, C.F. 1/743.

Table 8.9 Merchants who obtained more than three contracts

Name	Contracts obtained	Name	Contracts obtained
James Berry	12	Richard Martin	7
Georges Carter	7	Samuel Mellish	21
John Cook	10	William Metherell	5
Richard Cross	10	William Preddy	6
William Lea	4	Allen Spencer	14
Thomas Lewis	37	William Stevens	9
		Nathanael Walker	6
Contracts offered to these 13 merchants: 148			

Table 8.10 The victualling of the Royal Navy in London with salted meat

A – Merchants who took part in invitations to tender bids

***Merchants acting individually* 22**

James Adams	John and Peter Lefebvre	Daniel Ponton
Thomas Ayres	Isaac Lefebvre	Edmond Russel
Thomas Bell	Lloyd	Allen Spencer
James Berry	Thomas Lewis	H. Wilkes
Georges Deake	Jos. Mawbey	John Woodard
Edward Edes	Peter Mellish	Roger Wright
William Ford	Samuel Mellish	
T. Hurford	John Pass	

***Associations* 16**

Messrs Barnevelt and Thownshend	Messrs Harding and Ford
Messrs Bell and Grace	Messrs Lewis and Mellish
Messrs Bell and Lefebvre	Messrs Lewis, Spencer and Woodard
Messrs Bell and Mason	Messrs Lewis, and Spencer
Messrs Carter and Edes	Messrs Lewis and Woodard
Messrs Grace and Olier	Messrs Mawbey and Foster
Messrs Grace, Lloyd and Olier	Messrs Mellish and Preddy
Messrs Harding and Cartier	Messrs Woodard and Carter

B – A breakdown of the contracts awarded

1 – Breakdown of contracts for beef

***Number of contracts awarded* 12**

***Number of contracts obtained by individual merchants* 4**

James Berry: 1
Thomas Lewis: 1
Samuel Mellish: 1
Allen Spencer: 1

***Contracts obtained by associations* 8**

Messrs T. Lewis and S. Mellish: 1
Messrs T. Lewis, A. Spencer and J. Woodard: 1
Messrs T. Lewis and A. Spencer: 3
Messrs T. Lewis and J. Woodard: 1
Messrs S. Mellish and W. Preddy: 1
Messrs J. Woodard and G. Carter: 1

Individuals who participated in this victualling 6
(James Berry, Thomas Lewis, Samuel Mellish, William Preddy, Allen Spencer, John Woodard)

2 – Breakdown of contracts for pork

***Number of contracts awarded* 18**

***Number of contracts obtained by individual merchants* 17**

James Berry: 2
Daniel Bisson Junior: 1
John Cook: 10
John and Isaac Lefebvre: 2
Johnson Pasco: 2

***Contracts obtained by associations* 1**

Messrs Cost, Smith and Heathfield: 1

Table 8.11 Associations and alliance networks

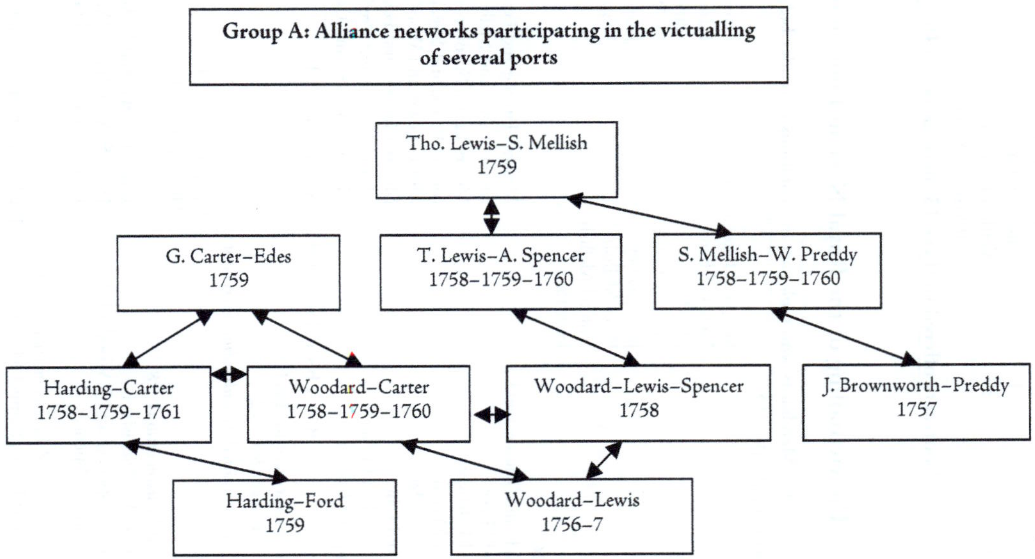

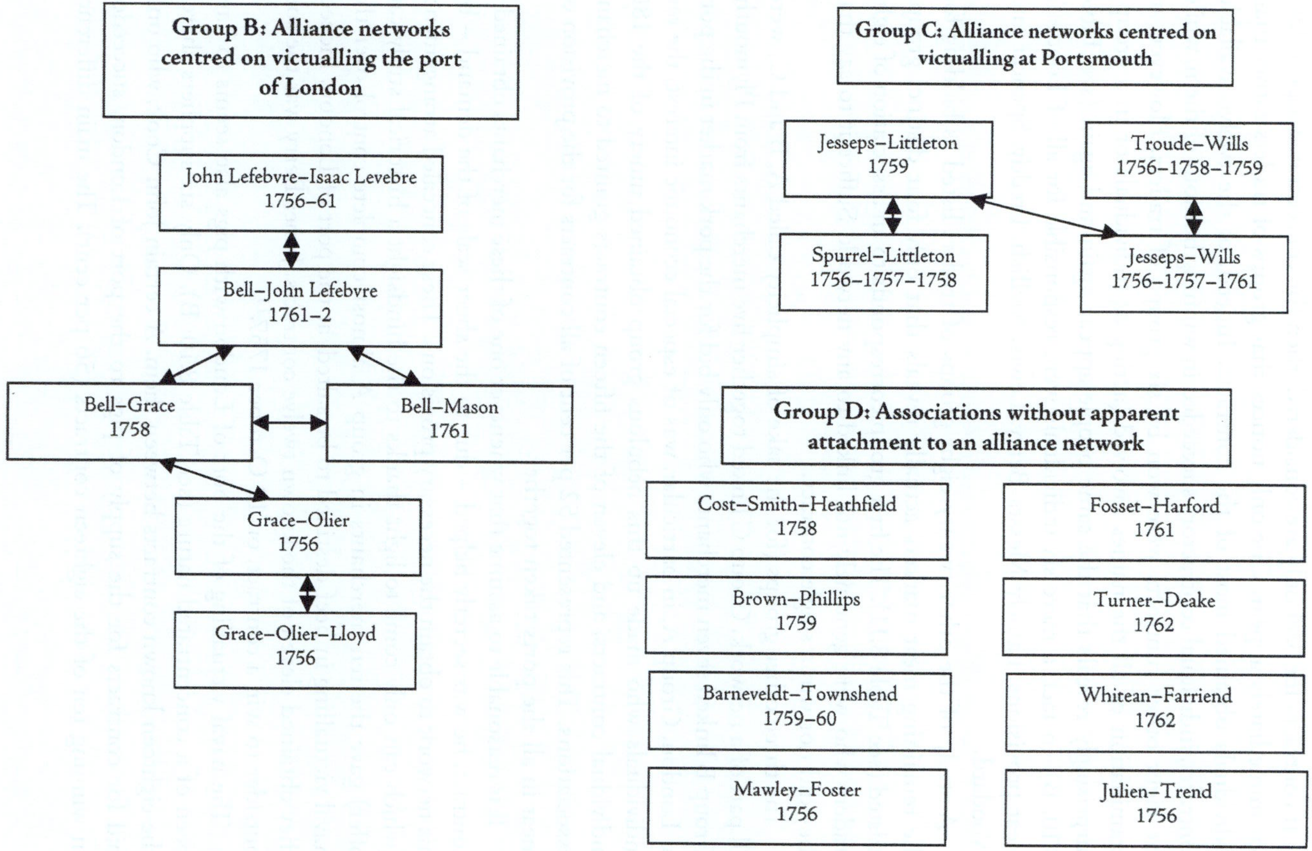
Group B: Alliance networks centred on victualling the port of London
John Lefebvre–Isaac Levebre 1756–61
Bell–John Lefebvre 1761–2
Bell–Grace 1758
Bell–Mason 1761
Grace–Olier 1756
Grace–Olier–Lloyd 1756
Group C: Alliance networks centred on victualling at Portsmouth
Jesseps–Littleton 1759
Troude–Wills 1756–1758–1759
Spurrel–Littleton 1756–1757–1758
Jesseps–Wills 1756–1757–1761
Group D: Associations without apparent attachment to an alliance network
Cost–Smith–Heathfield 1758
Fosset–Harford 1761
Brown–Phillips 1759
Turner–Deake 1762
Barneveldt–Townshend 1759–60
Whitean–Farriend 1762
Mawley–Foster 1756
Julien–Trend 1756

the London market, in spite of appearances, was concentrated in the extreme. Table 8.10: B illustrates this point by regrouping the data.

If contracts for beef only are considered, then the concentrated aspect does not immediately appear. One only notices that groups of traders rather than individuals obtained most of the contracts. Indeed, of the twelve available contracts, individual contractors succeeded in winning but four of them, while the eight other contracts were won by six groups of traders. However, an examination of all the names involved (acting as individuals or in a group) surprisingly reveals that the same people appear again and again (see Table 8.10: B). In fact, a mere six individuals were responsible for all of London's meat provisions, namely Messrs Berry, Lewis, Mellish, Preddy, Spencer and Woodard.

A study of the other twenty-eight groups of traders listed as bidding for the remaining meat markets actually reveals that only four distinct groups existed (see Table 8.11). The first group corresponds to an association of eight traders who were genuinely not linked to any network. Suffice it to say that they did not win a single contract.

The three other groups (for the sake of simplicity called A, B and C) were all part of a network. Group C linked together five merchants from Plymouth. Group B linked seven merchants who only bid for the pork market in the port of London. Group A, in particular, was of especial economic interest: the ten individuals who made up this nebulous group obtained ninety of the 180 individual contracts, and eleven of the fifteen contracts granted to merchant associations. This represented 52 per cent of all contracts for the provision of meat in all the ports taken together.

It is reasonable to assume that whenever one of these merchants obtained a contract, he was secretly helped – due to the sheer scale of the demand – by his network to obtain the necessary provisions. These concealed arrangements (which can only come to light thanks to the hindsight a historical study can offer) gave the ten merchants in group A almost complete control over the naval victualling in beef destined to be salted, at the port of London. Indeed, they obtained eleven of the known twelve contracts. James Berry was the sole outsider to win a contract, on 12 October 1757.[13]

The naval victualling of the port of London with pigs also seems to have been of a concentrated nature (see Table 8.10: B). Only six suppliers shared the eighteen known contracts between them. A certain John Cook, who only bid for contracts for the supply of pigs to the port of London, succeeded in winning ten of the eighteen contracts (56 per cent). The main difference

13 PRO, ADM 111/46, 12 October 1757.

between this market and the beef one is that only one contract was awarded to a group or association of merchants.

The remaining three ports (see Tables 8.12, 8.13 and 8.14) seemed to present a similar profile. It was unfortunately impossible to examine the pork and beef markets as distinct groups.

Table 8.12 The victualling of the Royal Navy in Portsmouth with salted meat

Merchants who took part in the invitations to tender bids		
***Merchants acting individually* 18**		
(-) (number of participations)		
William Colten (2)	William Lea (5)	William Preddy (1)
William Ford (1)	Thomas Lewis (2)	Thomas Purdery (1)
Robert Fowler Coade (1)	Peter Mellish (1)	Allen Spencer (7)
Joseph Fysson (3)	Samuel Mellish (3)	William Stevens (6)
Richard Joliffe (3)	Johnson Pasco (2)	Daniel Whitbean (1)
Samuel Johson (1)	James Perry (1)	Robert Willian (1)
***Associations* 0**		
Breakdown of the contracts awarded		
***Number of contracts awarded* 12**		
***Merchants* 6**		
(L): Local trading only		
William Colten: 1	Richard Joliffe: 2	
Robert Fowler Coade: 1 (L)	Allen Spencer: 3	
Joseph Fysson: 2	William Stevens: 3	

In Portsmouth, only six of the eighteen merchants who tendered bids for the twelve available contracts were successful. Allen Spencer and William Stevens were particularly noticeable because they won three contracts each and tendered bids at seven and six invitations to tender, respectively (see Table 8.12).

In Dover, local merchants who only operated locally were more striking than at Portsmouth (see Table 8.13). Seven of the twelve possible contracts were awarded to three local merchants, Matthew Kennet, Henry Sayer and Nathaniel Walker. The powerful merchants, men such as Allen Spencer, Thomas Lewis and William Preddy, only succeeded in winning one contract each.

In Plymouth, the domination of the market by local merchants was absolute, as is testified by the fact that all ten contracts were awarded to merchants who did not place bids at other ports (see Table 8.14). One merchant in particular stands out from the others: the Exeter butcher Richard Cross (close to Torbay), the only merchant to have obtained more than one contract (in his case, three), and to have bid at nine out of ten calls for tender. The Plymouth market seems to have been less concentrated than those of Dover and

Table 8.13 Merchants competing for the victualling of the Royal Navy in Dover with salted meat

Merchants who took part in the invitations to tender bids

***Merchants acting individually* 16**

(-) (number of participations)

James Berry (2)
Joseph Brownsworth (1)
Nicholas Burvit (2)
Georges Carter (1)
Martin Child (1)
Thomas Henley (1)
Matthew Kennet (1)
Thomas Lewis (4)
Samuel Mellish (5)
Edward Monns (1)
William Preddy (4)
Henry Sayer (3)
Allen Spencer (8)
Nathanael Walker (9)
Thomas Webb (4)
John Woodard (1)

***Associations* 1**

Messrs Woodard and Lewis (1)

Breakdown of the contracts awarded

Number of contracts awarded 12

Merchants 8

(L): Local trading only

James Berry: 1
Matthew Kennet: 1 (L)
Thomas Lewis: 1
William Preddy: 1
Henry Sayer: 3 (L)
Allen Spencer: 1
Nathanael Walker: 3 (L)
John Woodard: 1

Table 8.14 Merchants competing for the victualling of the Royal Navy in Plymouth with salted meat

Merchants who took part in the invitations to tender bids

***Merchants acting individually* 27**

(-) (number of participations)

John Chilcot (1)
William Clark (1)
Richard Cross (9)
John Culine (2)
Peter Culune (1)
Richard Disting (1)
William Earled (1)
Thomas Edmunds (1)
Thomas Edwards (4)
Joseph Fysson (1)
Francis Giles (1)
John Jesseps (1)
Thomas Jewell (1)
Richard Joliffe (1)
John Lear (2)
Edward Littleton (1)
John Pass (4)
Robert Philipps (2)
Richard Ridgeby (1)
John Snell (1)
John Spurrel (1)
William Stevens (1)
Thomas Symons (1)
John Troude (1)
Richard Turner (1)
John Wills (1)

***Associations* 5**

Messrs Brown and Philipps (1)
Messrs Jesseps and Littleton (1)
Messrs Jesseps and Wills (3)
Messrs Spurrel and Littleton (5)
Messrs Troude and Wills (5)

Breakdown of contracts awarded

***Number of contracts awarded* 10**

***Merchants* 8**

(L): Local trading only

Richard Cross: 3 (L)
John Culine: 1 (L)
Thomas Edwards: 1 (L)
Thomas Jewell: 1 (L)
Robert Philipps: 1 (L)
Thomas Symons: 1 (L)
John Troude: 1 (L)
Messrs Troude and Wills: 1 (L)

Table 8.15 Merchants competing for contracts for imported meat from Ireland

Merchants who took part in invitations to tender bids		
Merchants acting individually 13		
Anthony Bacon	James Evory	Henry Muilman
Barnevelt	P. Fearon	Arnold Nesbitt
James Benson	John Hallet	Francis Smith
John Biggin	John Ismay	
Richard Bradshaw	Mary Knowles	
Associations 1		
Messrs J. Diggs and Le Touch		
Breakdown of the contracts awarded		
Number of contracts awarded 19		
Merchants 7		
Anthony Bacon: 1	Henry Muilman: 1	
James Benson: 7	Arnold Nesbitt: 3	
James Evory: 4	Francis Smith: 2	
John Hallett: 1		

Portsmouth: eighteen and sixteen merchants in these two ports, respectively, participated in the invitation to tender bids, whereas in Plymouth twenty-seven individual merchants and five merchant associations were in competition.

In fact, it transpires that only three individual merchants were offered contracts in more than one of the four geographical zones under consideration: James Berry who was awarded four contracts (three contracts in London and one in Dover), Thomas Lewis who obtained two contracts (one in London and one in Dover), and Allen Spencer who succeeded in getting no less than five contracts in three zones (one in London, one in Dover and three in Portsmouth).

Finally, before proceeding to an examination of the victualling of these same ports in fresh meat, the final point to be considered in this section will be the supporting role provided by Ireland (Table 8.15). The Irish market was fairly well divided, with fourteen individual or group merchants in competition. However, only seven merchants actually obtained the nineteen available contracts, and three of them – James Benson, James Evory and Arnold Nesbitt – obtained fourteen of the nineteen contracts, representing 74 per cent of the total market share.

Suppliers of fresh meat

An examination of the body of fresh meat suppliers reveals that they operated in three distinct geographical areas: the southeast, Portsmouth, and finally Torbay, which was located in the same geographical zone as Plymouth.

It seems that it was possible to merge the four aspects of the southeast market in order to retain the same merchants (see Table 8.16). The four aspects were (1) the provision of cattle for the Thames, (2) the provision of cattle for the Downs, (3) the provision of mutton for Chatham, The Nore, Sheerness and Blackstakes, (4) the provision of mutton for Deptford or Gravesend for the preparation of portable soup.

Table 8.16 Merchants competing for the victualling of the Royal Navy with fresh meat in the southeast of England (Thames area, the Downs, Chatham, The Nore, Sheerness, Blackstakes, Deptford and Gravesend)

I – Fresh beef for the ships on the Thames	
***Number of contracts awarded* 26**	
***Suppliers* 5**	
James Berry: 1	Peter and Samuel Mellish: 7 (one with Lewis)
William Ford: 1	William Preddy: 2
Thomas Lewis: 15 (one with the Mellishes)	
II – Fresh beef for the Downs	
***Number of contracts awarded* 24**	
***Suppliers* 12**	
James Berry: 3	**Thomas Lewis:** 2
Joseph Brownsworth: 1	Samuel and Peter Mellish: 7
Ralph Chambers: 1	Messrs Preddy and Brownsworth: 1
Martin Child: 2	Allen Spencer: 3
Georges Deak: 1	Nathanael Walker: 2
William Ford: 1	
III – Wether mutton for ships anchored in Chatham or stationed off the coast at The Nore, Sheerness and Blackstakes	
***Number of contracts awarded* 27**	
***Suppliers* 9**	
Georges Carter: 7 (one with Lewis)	Richard Martin: 7
David Harridge: 1	Samuel Mellish: 2
Thomas Lewis: 9 (one with Carter)	Harris Roughton: 1
Messrs Lewis and Woodard: 1	Allen Spencer: 1
	John Woodard: 1
IV – Wether mutton with which to make portable soup	
***Number of contracts awarded* 9**	
***Suppliers* 3**	
Georges Carter: 1	Samuel Mellish: 1
Thomas Lewis: 7	

Nonetheless, one particular merchant held the most powerful position: Thomas Lewis, who was the contractor who had obtained the most contracts for the victualling of mutton for the preparation of portable soup for ships stationed in Chatham and The Nore (seven out of a possible nine contracts). He also obtained contracts for the provision of cattle to ships on the Thames (fifteen out of a possible twenty-six contracts). Only three other merchants provided minimal competition: Richard Martin and George Carter, who both obtained seven contracts for the provision of mutton to Chatham, and Samuel Mellish who obtained seven contracts for the provision of cattle to the Thames, and then succeeded in supplanting Lewis in the competition for the provision of cattle to the Downs by obtaining seven contracts to Lewis's two. This was probably due to Mellish being so well established in the Dover area.

The victualling of ships in Portsmouth with fresh meat created a different group of influential merchants. Although names such as Thomas Lewis and Samuel Mellish reappear, it is clear that this time they had a secondary role. Lewis, who participated in thirteen bids, only won one contract. On the other hand, Allen Spencer, William Stevens and James Berry, who were considered to be outsiders (based on their previous performance), succeeded in sharing 64 per cent of the contracts. Spencer and Stevens were simultaneously the main suppliers of salt beef to Portsmouth.

Table 8.17 Merchants competing for the victualling of the Royal Navy with fresh meat in Portsmouth

Number of contracts awarded 25	
***Suppliers* 9**	
James Berry: 5	Samuel Mellish: 3 (one with Preddy)
Edward Fossett: 1	William Preddy: 1 (one with Mellish)
Richard Joliffe: 1	Allen Spencer: 6 (one with Stevens)
William Lea: 4	William Stevens: 5 (one with Spencer)
Thomas Lewis: 1	

Table 8.18 The prices of fresh mutton and beef for ships anchoring in Torbay (in shillings/pence)

Bidding date	Price of mutton	Price of beef
21/05/1759	34/6	32/6
07/03/1760	30/10	28
04/06/1760	32/3	?
12/09/1760	28/4	23/9
16/03/1761	29/10	26/7
19/06/1761	25/4	23/6
23/09/1761	23/3	19/11
21/12/1761	22/5	24/3
29/03/1762	26/7	22/9
05/07/1762	25/10	23/3
29/06/1762	23/3	18/11

In the Torbay area, two merchants alone shared the eleven available contracts (see Table 8.19). Richard Cross of Exeter, this zone's most important supplier of salt beef, succeeded in obtaining seven of the eleven contracts, and the four remaining contracts went to William Metherall, a merchant from

Figure 8.6 Fluctuations in the prices of fresh meat (beef and mutton) destined for ships at anchor in Torbay

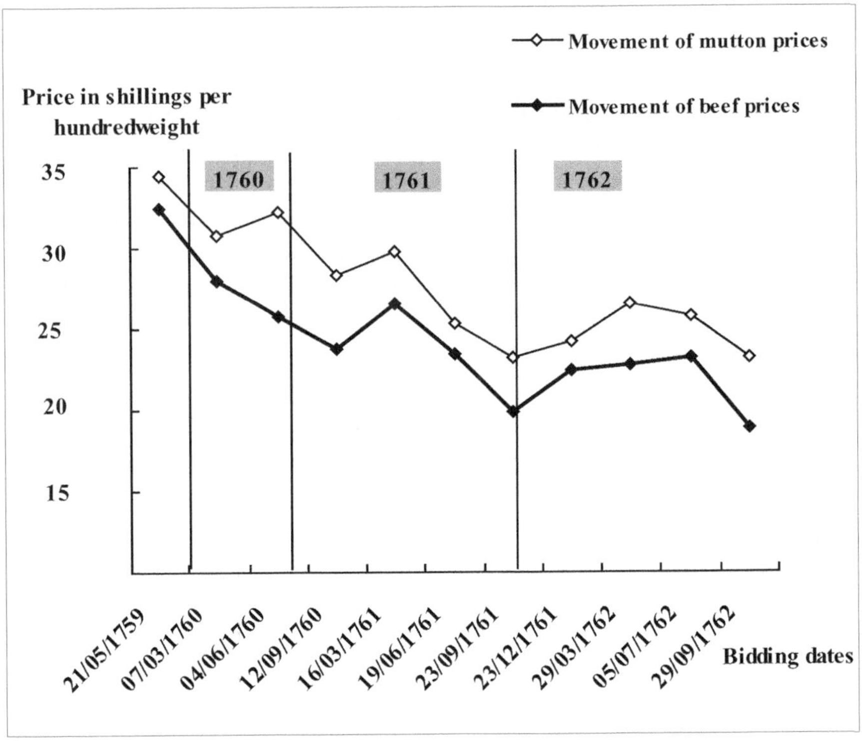

Paignton (a village bordering on Torbay), whose name appears for the first time.

The fact that a few other merchants participated in the bids for tender,[14] the significant decrease in prices throughout the bidding process (see Figure 8.6) and the open competition between Richard Cross and William Metherell suggest that, despite its highly concentrated profile, this segment of the market was also extremely competitive.

Table 8.19 Merchants competing for the victualling of Royal Navy shipsat anchor in Torbay with fresh meat

Number of contracts awarded 11
Suppliers 2
Richard Cross: 7 William Metherell: 4

14 John Troude, William Lake, Peter Mellish, Edward Brown and Richard Windsor.

*

In conclusion, it would appear that only the sector of activity involving the victualling of the port of London with beef destined for salting had a highly concentrated profile. The existence of a network whose members obtained eleven of the twelve contracts available explains why prices dropped so little and so infrequently compared to other meat victualling.

However, this phenomenon only had short-lived consequences, since the import of meat from Ireland from the spring of 1758 allowed the Commissioners of the Victualling Board to reconstruct the conditions necessary for a competitive market by widening the bids base, which in turn would result in a substantial and continuous decrease in prices, as was the case for other types of contracts.

9

Cereals and Pulses

This chapter, which is devoted to the category of cereals and pulses, will examine a group of products that was also often given the generalised label of '*sundry provisions*'. This will include wheat and its derivatives (flour and biscuits), oats, groats, malt and peas. The list was not drawn up according to whether a cereal or a pulse belong to a certain genus – for example, hops are not considered in this chapter – but because the items on the above list were all provided by a single body of merchants, consisting of eighty-four individual suppliers and five merchant associations, resulting in a headcount of eighty-eight individuals.

A greater understanding of this sector will be provided by an examination of the evolution of each market in terms of prices, followed by an analysis of how many merchants were involved in each market and to what extent real competition prevailed.

The Evolution of Different Segments of the Market in Terms of Prices and Quantities

Wheat, flour and biscuits

Between 21 April 1756 and 24 November 1762 the Commissioners in London organised no fewer than 104 auctions in order to provide the Navy with wheat – eighty-eight contracts were destined for the victualling of London, sixteen others for Plymouth.

Figure 9.1 illustrates the various parameters operating in terms of the eighty-eight London contracts. Wheat was bought on a monthly basis. Each delivery was for a pre-established quantity, usually of 2,400 quarters, and had to be delivered '*in a month by weekly proportion*', i.e. one quarter of the quantity was due every week. Only nine exceptions to this rule have been found, and they were the result of an increase in prices that led the Commissioners to reduce the quantities they had ordered. This occurred between August and November 1756, on 31 August and 10 October 1757. Since the bakehouses needed a continuous supply of flour, the Commissioners reacted by introducing new contracts on a fortnightly basis (see Figure 9.1).[1]

1 PRO, ADM 30/44, 258.

Wheat was delivered by the bushel for the price of £59 (£60 including the sack) to Peace's Mills near Deptford or to the King's Mills in Rotherhithe. Each contract stated the exact proportion of wheat that should be delivered to the two sites (usually on a 50–50 basis).

An examination of Figure 9.1 reveals that the price fluctuation curve confirms W. Beveridge's suggestions concerning seasonal variation in price.[2] Prices generally peaked at the beginning of the spring (March, April and May) and reached their lowest level in the autumn. The transition between high and low prices was usually quite pronounced between the first half of the year (when they were high), and after the harvest (when they dropped). The years 1757–9 are indicative of this trend. The opposite movement in 1756 and 1761 does not invalidate this hypothesis, but was the consequence of a temporary trend that will also be examined in this chapter.

Figure 9.1 can be broken down into three main segments. The first segment covers a period running from April 1756 to July 1757 and can be characterised by an exponential rise in price. In May 1757, a quarter of wheat was worth 65 shillings, as against the rate of 29 shillings and 9 pence on 21 April 1756, which represents an increase of over 115 per cent in one year. This explosion in prices was the result of a combination of the bad harvest of 1756[3] and the Navy's increased demand following its growth in numbers (see Figure 8.4).

Then, from August 1757, there followed a four-year period in which prices took a downward turn. This tendency cannot be accounted for by a decrease in demand, since, on the contrary, the demand increased: the Navy now required not 2,000 but 2,400 and sometimes 3,000 quarters of wheat per month to satisfy its numbers which were still growing. This reduction, which followed the trend of many other products in the 1758–9 period, was undoubtedly due to savings the Royal Navy was now able to make by being able to pay off its debts in the allotted time (see Chapter 3).

Finally, the third segment begins in August 1761, when another period marked by price increases begins, although this one seems to have no common points with the first trend (see Figure 9.1 in which the highest price in this third segment reached 37 shillings and 7 pence). The characteristics of this segment were a product of the Navy's increasing difficulty in settling with its creditors by the required dates, and of the poor harvest in 1761 – discussed by T.S. Ashston,[4] as well as J.D. Chambers and G.E. Mingay.[5]

2 Beveridge, *Prices and Wages*, 536.

3 Macdonald, 'A new myth of naval history?', 174–5.

4 T.S. Ashton, *Economic Fluctuations in England, 1700–1800* (Oxford, Clarendon Press, 1959), 21, 36–8.

5 J.D. Chambers and G.E. Mingay, *The Agricultural Revolution in England 1750–1880* (London, Batsford, 1956), 43.

Figure 9.1 Fluctuations in prices and quantities of wheat bought for the port of London

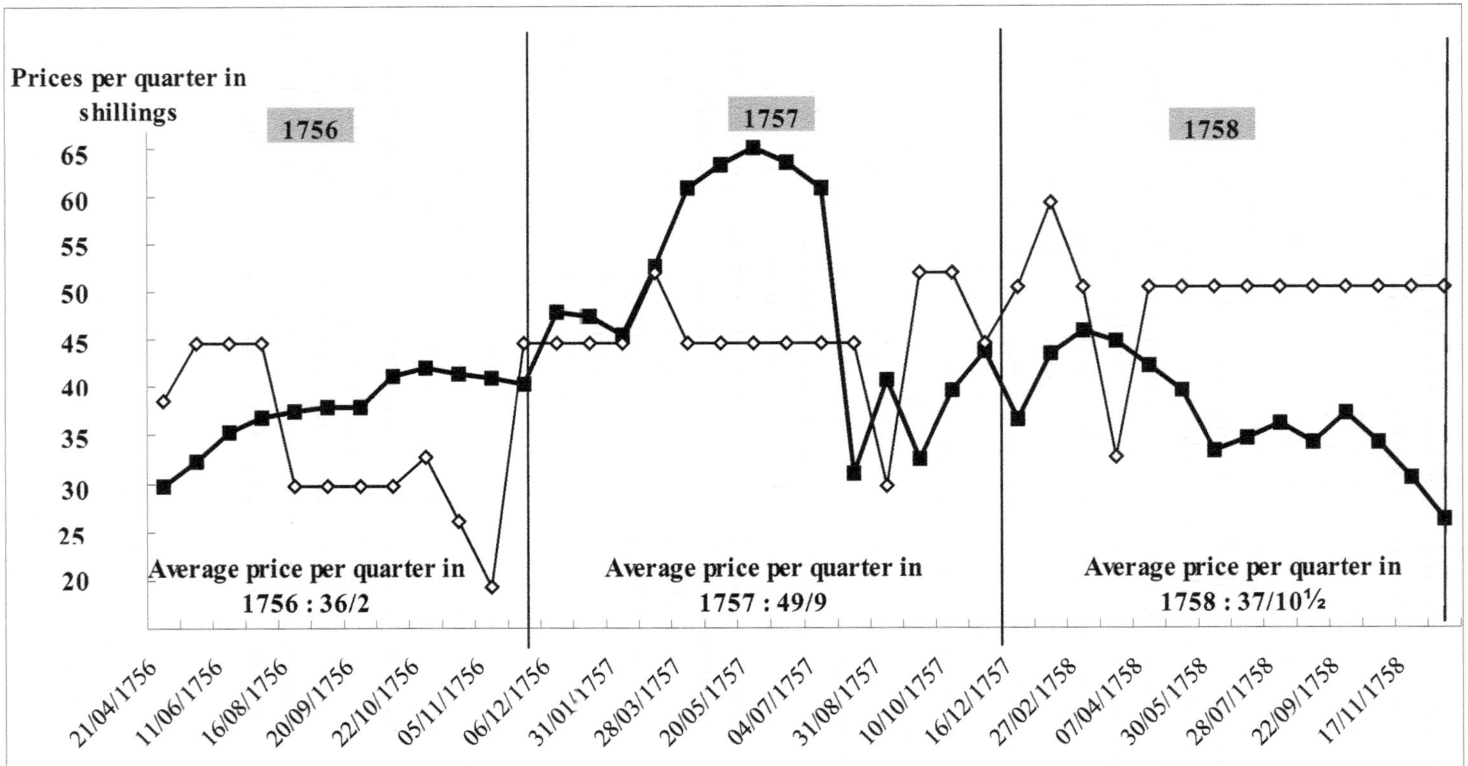

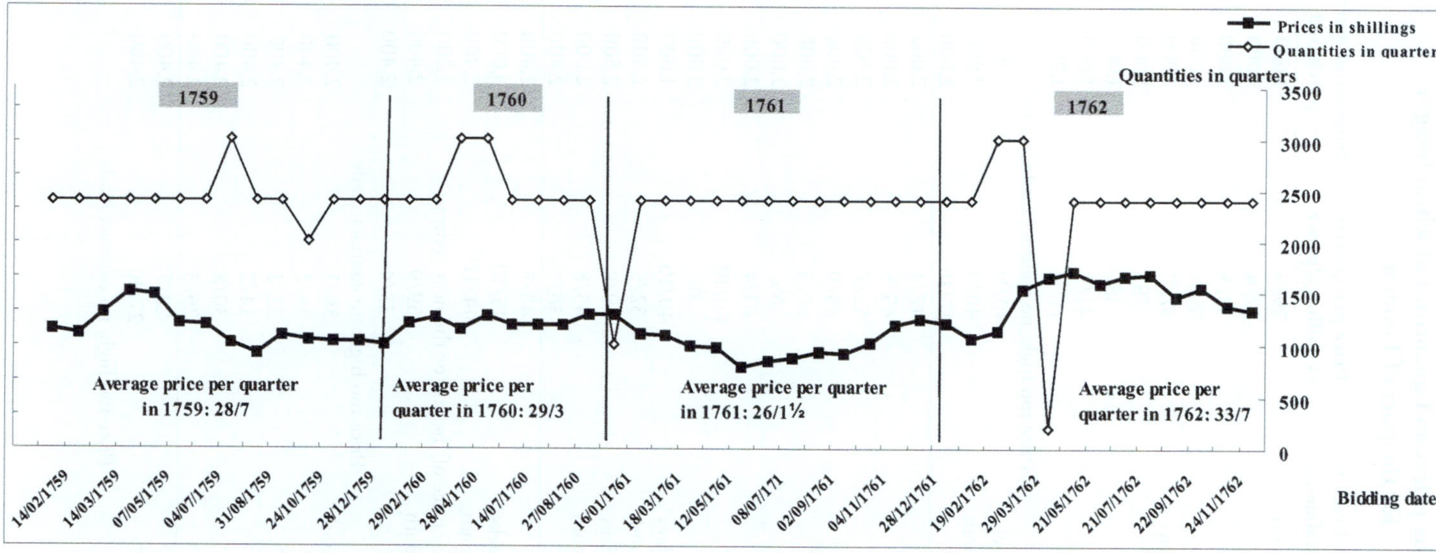
Prices in shillings
Quantities in quarters
Quantities in quarters
1759
1760
1761
1762
3500
3000
2500
2000
1500
1000
500
0
Average price per quarter in 1759: 28/7
Average price per quarter in 1760: 29/3
Average price per quarter in 1761: 26/1½
Average price per quarter in 1762: 33/7
14/02/1759
14/03/1759
07/05/1759
04/07/1759
31/08/1759
24/10/1759
28/12/1759
29/02/1760
28/04/1760
14/07/1760
27/08/1760
16/01/1761
18/03/1761
12/05/1761
08/07/171
02/09/1761
04/11/1761
28/12/1761
19/02/1762
29/03/1762
21/05/1762
21/07/1762
22/09/1762
24/11/1762
Bidding dates

Table 9.1 The prices and quantities of wheat bought for the port of London

Bidding date	Period of delivery by weekly instalments	Price per quarter in shillings/pence	Number of quarters purchased
21/04/1756	1 month	29/9	1,600
10/05/1756	"	32/4	2,000
11/06/1756	"	33/5	2,000
16/07/1756	"	36/10	2,000
16/08/1756	15 days	37/7	1,000
06/09/1756	"	38	1,000
20/09/1756	"	38	1,000
08/10/1756	"	41/3	1,000
22/10/1756	"	41/11	1,200
01/11/1756		Price too high: no contract made	
05/11/1756	1 week	40/11	300
12/11/1756	1 month	40/4	2,000
06/12/1756	"	47/10	2,000
03/01/1757	"	47/4	2,000
31/01/1757	"	45/5	2,000
28/02/1757	"	52/7	2,500
28/03/1757	"	69/9	2,000
24/04/1757	"	60/3	2,000
20/05/1757	"	65	2,000
15/07/1757	"	63/4	2,000
04/07/1757	"	60/10	2,000
05/08/1757	"	31	2,000
31/08/1757	15 days	40/10	1,000
09/09/1757	1 month	32/5	2,500
10/10/1757	15 days	39/8	2,500
21/11/1757	1 month	43/8	2,000
16/12/1757	"	36/7	2,400
11/01/1758	"	43/7	2,400
27/02/1758	5 weeks	45/10	3,000
29/03/1758	1 month	44/11	2,400
07/04/1758	Offer at the price of the last or the next contract		1,200
01/05/1758	1 month	39/9	2,400
30/05/1758	"	33/4	2,400
30/06/1758		Price too high: no contract made	
28/07/1758	1 month	36/3	2,400
25/08/1758	"	34/2	2,400
22/09/1758	"	37/2	2,400
25/10/1758	"	34/2	2,400
17/11/1758	"	30/8	2,400
15/12/1758	"	26/3	2,400
12/01/1759	"	28/7	2,400
14/02/1759	"	27/10	2,400
09/03/1759		Price too high: no contract made	

14/03/1759	1 month	34/5	2,400
16/03/1759	"	34	2,400
07/05/1759	"	29/6	2,400
06/06/1759	"	29/4	2,400
04/07/1759	"	26/5	3,000
06/08/1759	"	24/11	2,400
31/08/1759	"	27/7	2,400
26/09/1759	"	26/11	2,000
24/10/1759	"	26/10	2,400
23/11/1759	"	26/8	2,400
28/12/1759	"	26/3	2,400
04/02/1760	1 month	29/7	2,400
29/02/1760	"	30/4	2,400
28/03/1760	"	28/8	2,400
28/04/1760	"	30/9	2,400
02/06/1760	"	29/4	2,400
14/07/1760	Offer at the price of the last or the next contract		
25/07/1760	1 month	29/2	2,400
27/08/1760	"	30/10	2,400
22/09/1760	"	30/10	2,400
16/01/1761	"	27/10	2,400
20/02/1761	"	27/7	2,400
18/03/1761	"	26	2,400
15/04/1761	"	25/10	2,400
12/05/1761	"	22/9	2,400
12/06/1761	"	23/8	2,400
08/07/1761	"	24/2	2,400
05/08/1761	"	25	2,400
02/09/1761	"	24/9	2,400
05/10/1761	"	26/5	2,400
04/11/1761	"	29/3	2,400
02/12/1761	"	30/2	2,400
28/12/1761	"	29/5	2,400
27/01/1762	"	27/3	2,400
91/02/1762	"	28/3	3,000
19/03/1762	"	34/9	3,000
29/03/1762	Offer	36/9	200
19/04/1762	1 month	37/7	2,400
21/05/1762	"	35/10	2,400
21/06/1762	"	36/10	2,400
21/07/1762	"	37/2	2,400
23/08/1762	"	33/7	2,400
22/09/1762	"	35/2	2,400
18/10/1762	"	32/4	2,400
24/11/1762	"	31/6	2,400

Before an analysis of wheat provisions for Plymouth purchased in London can be made, the elasticity between price variation and quantity should be borne in mind (see Figure 9.1). The gap of one month that can sometimes be noticed between a price fluctuation and the opposite trend in quantity results from the fact that the Commissioners would base the quantities they requested on the price of the last contract which had been announced at the last invitation to tender bids and which had appeared in the press. The Commissioners were thus able to lessen any big discrepancies between offer and demand and, conversely, build up some stocks during less tense periods.

The Commissioners contracted agreements for the delivery of quarters of wheat to Plymouth on sixteen occasions, contrasting with only one such agreement for both Portsmouth and Dover.[6] However, these contracts would fade out over time: five contracts were issued in 1756; four in 1757; one in 1758, 1759 and 1760; and two in 1761 and 1762 (see Figure 9.2). These must be considered as attempts to ease pressure on the London mill during a period of great tension, since it should be remembered that, having been bought in London, wheat, transformed into flour, was regularly dispatched to Dover, Portsmouth and Plymouth (see Chapters 2 and 4).

Figure 9.2 Fluctuations in prices and quantities of wheat bought for the port of Plymouth

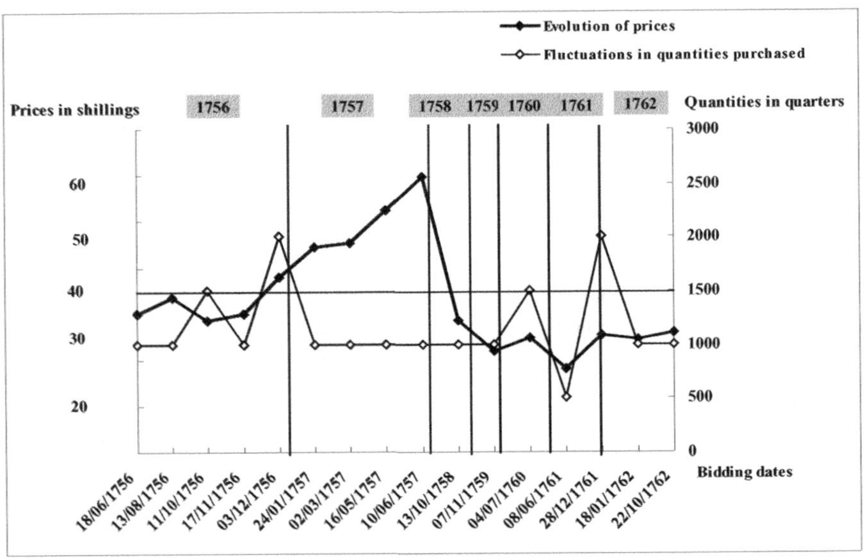

6 On 2 December 1757 for Dover and on 8 June 1762 for Portsmouth. PRO, ADM 111/46 and 51.

The contractors' tenders remained unchanged throughout the duration of the war. Once a price had been agreed upon per quarter, the contractors were committed to delivering wheat bought in the king's warehouses in Plymouth '*free of all charge and risk*'. The only slight variation to occur over the years concerned the deadlines in loading the produce: in 1756, suppliers were granted ten to twelve days; as from 1757, a fortnight was generally allocated and their supply ship had '*to sail afterwards as soon as wind and weather permits*'. A copy of each contract was sent to the Agent in Plymouth. These deliveries, similarly to London, were carried out on the basis of sacks containing 59 pounds of wheat.

Despite having been bought on the London market, wheat destined for Plymouth also underwent similar price fluctuations to those illustrated in Figure 9.1 (cf. Figure 9.2). There was also an increase of four to five shillings a quarter which was the result of supplementary costs that the suppliers were responsible for.

Table 9.2 The price and quantities of wheat bought for Plymouth

Bidding date	Period of delivery by weekly instalments	Price per quarter in shillings	Number of quarters purchased
18/06/1756	10 days	39/11	1,000
13/08/1756	12 days	43/6	1,000
11/10/1756	10 days	38/8	1,500
17/11/1756	14 days	40	1,000
03/12/1756	3 weeks	48	2,000
24/01/1757	14 days	54/6	1,000
02/03/1757	14 days	55/5	1,000
16/05/1757	Price too high: no contract made		
10/06/1757	14 days	69/8	1,000
13/10/1758	14 days	38/5	1,000
07/11/1759	10 days	31/8	1,000
04/07/1760	10 days	34/7	1,500
08/06/1761	14 days	27/10	500
28/12/1761	3 weeks	35/5	2,000
18/01/1762	Offer (immediately)	34/5	1,000
22/10/1762	10 days	35/10	1,000

Purchases of flour and biscuits, unlike purchases of wheat for London, were not based on a fixed rhythm. They occurred when the Victualling Board's flour mills and/or bakehouses were working at full capacity and yet were still unable to meet the demand. In these cases, semi-finished products such as flour, or finished products such as biscuits were bought to avoid bottlenecks. This situation reflected the Victualling Board's incapacity sometimes to meet all its demands, and accounted for the irregularities apparent in its contracts in terms of duration and in terms of quantity.

Figure 9.3 Fluctuations in the prices of flour and quantities bought in London by the Commissioners of the Victualling Board

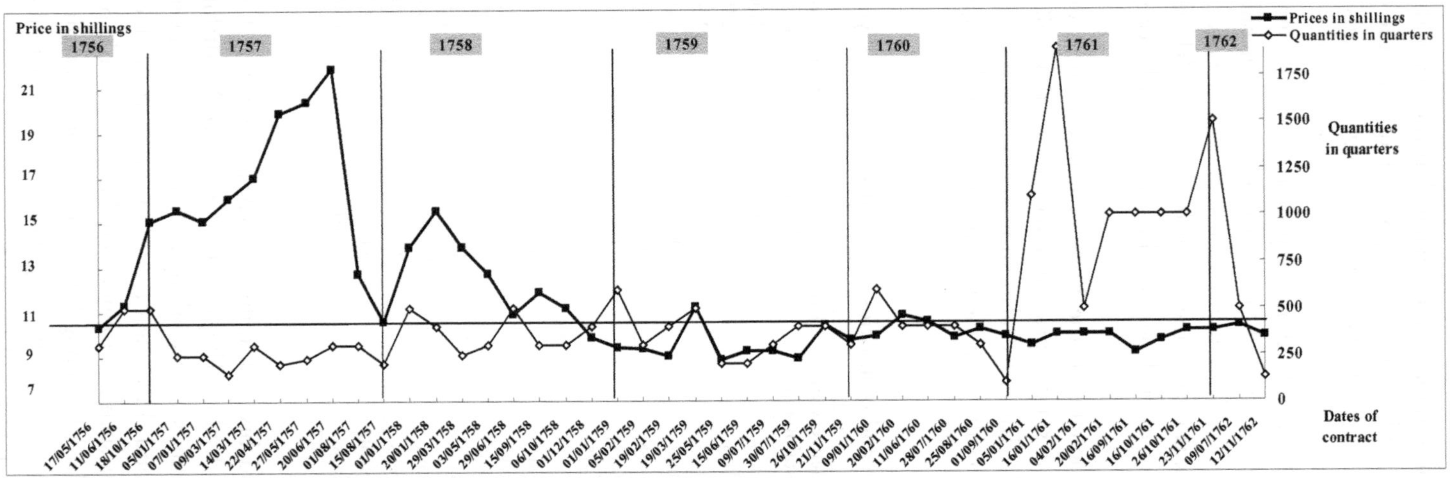

Figure 9.4 Fluctuations in prices and quantities of biscuits bought in London by the Victualling Board

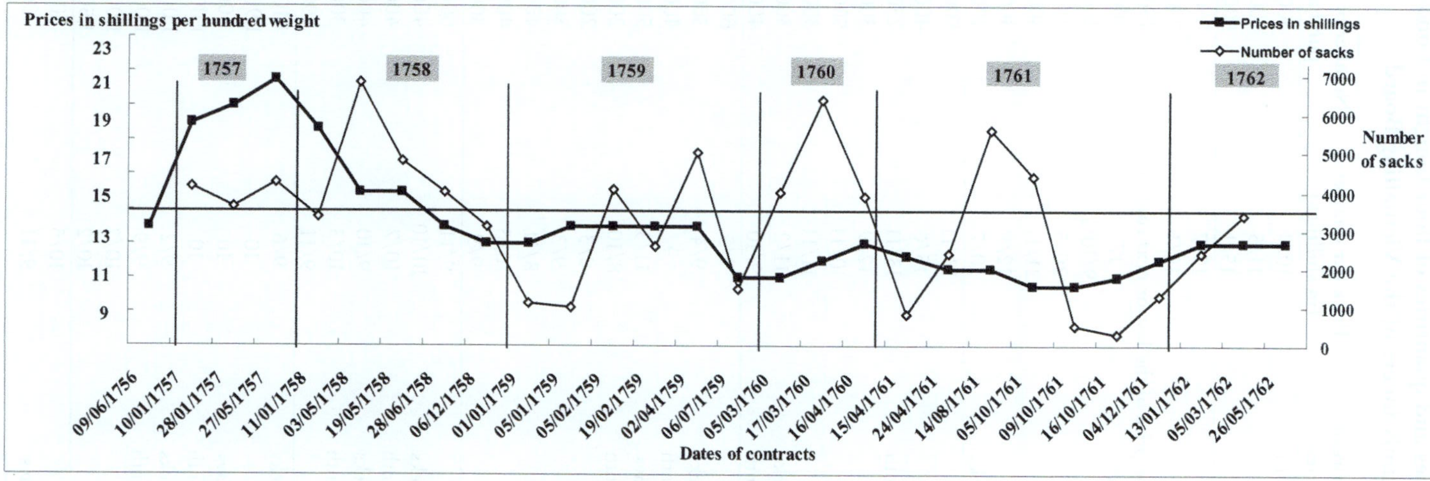

Table 9.3 The prices and quantities of flour bought in London by the Commissioners of the Victualling Board

Bidding date	Time between deliveries	Price per quarter (in shillings)	Number of quarters purchased
17/05/1756	1 month	10/4	300
11/06/1756	"	11/5	500
18/10/1756	"	15/1	500
05/01/1757	"	15/7	250
07/01/1757	"	15/1	250
09/03/1757	Offer at the price of the last or next contract		150
14/03/1757	15 days	17	300
22/04/1757	"	19/10½	200
27/05/1757	"	20/5	230
20/06/1757	"	21/11	300
01/08/1757	"	12/9	300
15/08/1757	3 weeks	10/7	200
01/01/1758	"	13/11	500
20/01/1758	"	15/6	400
29/03/1758	1 month	13/11	250
03/05/1758	"	12/9	300
29/06/1758	"	10/11	500
15/09/1758	"	11/11	300
06/10/1758	3 weeks	11/2	300
01/12/1758	1 month	9/10	400
01/01/1759	"	9/5	600
05/02/1759	3 weeks	9/4	300
19/02/1759	1 month	9	400
19/03/1759	15 days	11/3	500
25/05/1759	1 month	8/10	200
16/06/1759	"	9/3	200
09/07/1759	"	9/3	300
30/07/1759	"	8/11	400
26/10/1759	"	10/6	400
21/11/1759	"	9/9	300
09/01/1760	"	9/11	600
20/02/1760	3 weeks	10/10	400
11/06/1760	1 month	10/7	400
26/07/1760	3 weeks	9/10	400
25/08/1760	1 month	10/3	300
01/09/1760	"	9/11	100
05/01/1761	3 weeks	9/6	1,100
16/01/1761	"	10	1,900
04/02/1761	14 days	10	500
20/02/1761	1 month	10	1,000
16/09/1761	6 weeks	9/2	1,000
16/10/1761	1 month	9/9	1,000
26/10/1761	"	10/2	1,000
23/11/1761	"	10/2	1,500
09/07/1762	"	10/5	500
12/11/1762	8 days	9/11	130

Table 9.4 The prices and quantities of biscuits bought in London by the Victualling Board

Bidding date	Number of merchants tendering	Price per quarter (in shillings)	Number of sacks
09/06/1756	14	14	?
10/01/1757	13	20	4,140
28/01/1757	15	21	3,600
27/05/1757	15	22/6	4,250
11/01/1758	14	?	3,350
03/05/1758	17	16	6,810
19/05/1758	15	16	4,800
28/06/1758	18	14	4,000
06/12/1758	12	13	3,090
01/01/1759	11	13	1,100
05/01/1759	7	14	1,010
05/02/1759	18	14	4,550
19/02/1759	13	14	2,580
02/04/1759	18	14	5,010
06/07/1759	28	11	1,480
05/03/1760	17	11	3,970
17/03/1760	18	12	6,360
16/04/1760	23	13	3,867
15/04/1761	22	12/3	800
24/04/1761	?	11/6	2,400
14/08/1761	20	11/6	5,600
05/10/1761	?	10/6	4,400
09/10/1761	1 (offered)	10/6	500
16/10/1761	19	11	500
04/12/1761	Price too high (12/6): no contract made		
13/01/1762	12	13	2,400
05/03/1762	11	13	3,400
26/05/1762	Price too high (12): no contract made		

The adding together of yearly purchases of flour nonetheless shows a steady and constant increase in volume: 2,180 quarters in 1757, 2,950 quarters in 1758 and 3,600 quarters in 1759, culminating in 9,000 quarters in 1761. The figure for 1760 cannot be calculated since documents are missing in the archives for the second half of the year.

If the increased quantities noted between 1757 and 1759 can be accounted for as the direct result of a parallel increase in numbers, the explosion of 1761 needs to be accounted for more precisely. It seems to have resulted from a shortage in the Victualling Board's flour mills, coupled with an increase in demand following a fire that destroyed one of the mills in Deptford.[7] An analysis of the price curve in Figure 9.3 reveals that the movement was clearly the

7 PRO, ADM 110/20, 27 February 1760.

result of competitive prices in 1761 for flour as opposed to those for wheat. In fact, wheat prices had risen as a consequence of the poor harvest of 1761, whereas wheat flour milled from previous years' harvests remained stable in terms of price. Consequently, the Commissioners thought it wiser to level out purchases of wheat in spite of the increasing demand, and to concentrate on purchases of flour.

The volume of sacks of biscuits purchased followed a totally different pattern. Having reached an all-time high in 1758 with the purchase of 22,050 sacks, purchases then dropped to 15,730 sacks in 1759 and 14,000 sacks in 1761. This offers another insight towards understanding the parallel increase in the purchase of flour. In fact, it is more than likely that improvements in the processing of flour into biscuits or into bread in the bakehouses of the Victualling Board during the course of the war (cf. Chapter 4) resulted in diminished quantities of purchased biscuits.

Fluctuations in the price of wheat eventually mirrored decreased prices for flour and biscuits over a number of years (cf. Figures 9.3 and 9.4), with a 50 per cent price decrease for wheat, flour and biscuits in relation to the record price obtained in 1757 and the minimum price obtained in 1761.

Oats and groats

The acquisition of oats and groats – a semi-finished product originating from the milling of oats into flour – was carried out by Commissioners of the Victualling Board. An examination of Figures 9.5 and 9.6 reveals the Commissioners' decision in 1760 to stop all purchases of oats in favour of the purchase of groats. Consequently, the average volume of groats purchased in a year – between 3,000 and 4,000 quarters (3,800 quarters in 1757, 4,000 quarters in 1758, 3,150 quarters in 1759) – suddenly jumped to 9,300 quarters in 1761.

This reorientation of victualling practices confirms the pattern that has already been highlighted for wheat. Once again, it was a consequence of the Victualling Board's London mill's incapacity to process more than 33,000 quarters per annum. In fact, the adding together of wheat and oat purchases for milling into flour reveals that the volume remained constant year after year, thus highlighting maximal production capacity in the Deptford and Rotherhithe mills (31,700 quarters in 1757, 30,600 quarters in 1758, 32,900 quarters in 1759, 31,200 quarters in 1761). It must not be forgotten that a mill was destroyed by fire in February 1760, and that the other mills simply could not produce more flour. Since an ever-increasing number of servicemen needed more and more bread and biscuits, and this involved an increase in purchases of wheat, the Commissioners compensated by stopping the purchase of oats, and began to directly buy semi-finished products such as flour and groats.

Figures 9.5 and 9.6 also reveal the irregularity in dates for contracts issued for oats and groats. Delivery dates, on the other hand, were fairly regular

Table 9.5 The prices and quantities of oats bought in London by the Commissioners of the Victualling Board

Bidding date	Period of delivery by weekly instalments	Price per quarter (in shillings)	Number of sacks
05/05/1756	1 month	13/10	500
28/07/1756	14 days	18/8	100
11/08/1756	1 week	20	100
18/08/1756	15 days	19/6	200
08/09/1756	Immediately	20	100
20/09/1756	1 month	21	500
04/10/1756	Immediately	21	100
18/10/1756	"	20/9	100
27/10/1756	Price too high: no contract made		
01/11/1756	Price too high: no contract made		
24/11/1756	1 month	22	500
17/01/1757	15 days	23	500
04/02/1757	Offer at price of the last or next contract		200
30/03/1757	1 month	22/11	500
09/05/1757	"	24/5	500
04/07/1757	Price too high: no contract made		
13/07/1757	15 days	24	200
01/08/1757	1 month	19/4	300
29/08/1757	3 weeks	21/6	300
26/09/1757	"	23/5	300
20/02/1758	1 month	21/3	500
17/04/1758	"	19/9	400
24/05/1758	"	19/3	500
12/07/1758	"	16/2	500
24/11/1758	"	14/9	500
26/02/1759	"	14/4	500
30/04/1759	"	15/5	500
13/07/1759	"	12/5	500
04/06/1760	"	14/10	300

and were usually asked for '*in a month by weekly proportion*'. Price curves shown in the graphs are analogous to those for wheat, flour and biscuits. After an increase in prices in 1757, oat prices also began to decrease, eventually reaching a low of 50 per cent less than the highest price. Maximum and minimum prices for oats were 24s 6d in May 1757 and 12s 5d in July 1759; for groats they were 71s 5d in May 1757 and 33s in June 1760. There is, however, one variation to be found for groats in relation to the curves drawn for the prices for wheat and flour: while wheat and flour reached their lowest levels in 1761 (cf. Figures 9.1, 9.2 and 9.3), groat prices reached their lowest levels in June 1760. Their increase after this date was due to the large quantities of groats that the Commissioners had to buy from then on in order to

Figure 9.5 Fluctuations in prices and quantities of oats bought in London by the Commissioners of the Victualling Board

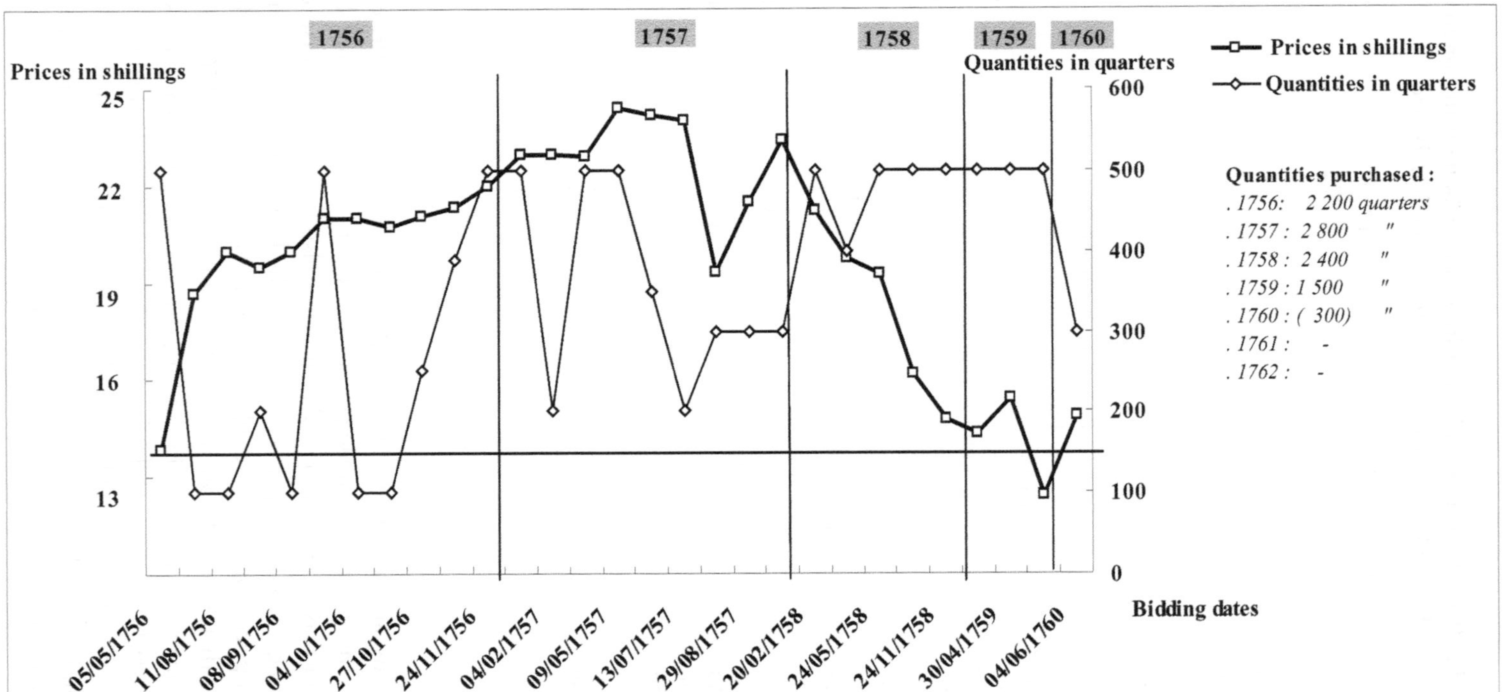

Figure 9.6 Fluctuations in prices and quantities of groats bought in London by the Commissioners of the Victualling Board

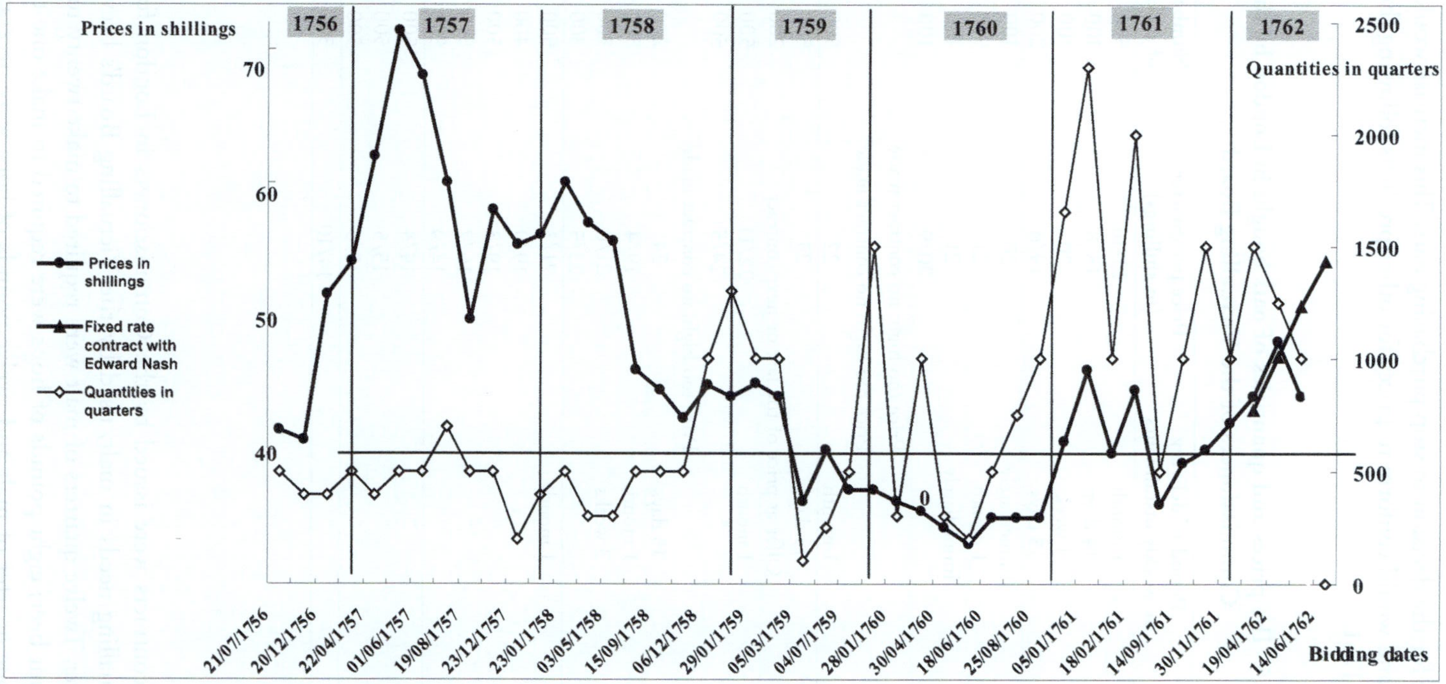

compensate for the decision to stop purchasing oats. This increase would end in 1761–2, and would conform to prices for other products following the bad harvests of 1761.

Table 9.5 The prices and quantities of oats bought in London by the Commissioners of the Victualling Board

Bidding date	Period of delivery by weekly instalments	Price per quarter (in shillings)	Number of sacks
05/05/1756	1 month	13/10	500
28/07/1756	14 days	18/8	100
11/08/1756	1 week	20	100
18/08/1756	15 days	19/6	200
08/09/1756	Immediately	20	100
20/09/1756	1 month	21	500
04/10/1756	Immediately	21	100
18/10/1756	"	20/9	100
27/10/1756	Price too high: no contract made		
01/11/1756	Price too high: no contract made		
24/11/1756	1 month	22	500
17/01/1757	15 days	23	500
04/02/1757	Offer at price of the last or next contract		200
30/03/1757	1 month	22/11	500
09/05/1757	"	24/5	500
04/07/1757	Price too high: no contract made		
13/07/1757	15 days	24	200
01/08/1757	1 month	19/4	300
29/08/1757	3 weeks	21/6	300
26/09/1757	"	23/5	300
20/02/1758	1 month	21/3	500
17/04/1758	"	19/9	400
24/05/1758	"	19/3	500
12/07/1758	"	16/2	500
24/11/1758	"	14/9	500
26/02/1759	"	14/4	500
30/04/1759	"	15/5	500
13/07/1759	"	12/5	500
04/06/1760	"	14/10	300

Malt

Forty-two contracts were issued by the Commissioners in London for the Navy's victualling needs in malt, used by the Victualling Board's breweries to make beer. Twelve quarters of malt were required to make twenty tons of petty warrant beer; eight pounds of hops were required to make one ton of beer; and one quarter of malt and twelve pounds of hops were necessary to

Table 9.6 The prices and quantities of groats bought in London by the Commissioners of the Victualling Board

Bidding date	Period of delivery by weekly instalments	Price per quarter (in shillings)	Number of quarters purchased
02/07/1756	1 month	41/6	500
23/08/1756	"	40/10	400
20/12/1756	"	51/6	400
04/03/1757	"	54	500
22/04/1757	"	62	400
27/05/1757	"	71/5	500
01/06/1757	3 weeks	68	500
25/07/1757	"	60	700
19/08/1757	15 days	49/9	500
25/11/1757	1 month	57/11	500
20/12/1757	15 days	55/3	200
09/01/1758	3 weeks	56	400
23/01/1758	"	60	500
21/03/1758	1 month	56/10	300
03/05/1758	"	55/6	300
17/07/1758	"	46	500
15/09/1758	"	44/6	500
01/11/1758	"	42/5	500
06/12/1758	"	44/11	1,000
01/01/1759	"	44	1,300
01/01/1759	"	45	1,000
01/01/1759	100 days	44	1,000
01/01/1759	1 month	36/2	100
01/01/1759	"	40	250
01/01/1759	"	37	500
25/01/1760	"	37	1,500
28/01/1760	Offer at price of last or next contract		300
24/03/1760	1 month	35/6	1,000
30/04/1760	Offer at price of last or next contract		300
04/06/1760	1 month	33	200
18/06/1760	"	34/11	500
22/08/1760		Price too high: no contract made	
25/08/1760	"	34/11	1,000
02/01/1761		Price too high: no contract made	
05/01/1761	3 weeks	46	2,300
02/02/1761	Offered	39/9½	1,000
18/02/1761	1 month	44/6	2,000
01/07/1761	"	36	500
14/09/1761	"	39	1,000
28/10/1761	"	40	1,500
30/11/1761	"	42	1,000
15/01/1762	1 week	44	1,500
	50 quarters a week	43	Contract at agreed price with Edward Nash
19/04/1762	1 week	48	1,000
	50 quarters a week	47	Contract at agreed price with Edward Nash
02/06/1762		Price too high: no contract made	
14/96/1762	50 quarters a week	54	Contract at agreed price with Edward Nash

Figure 9.7 Fluctuations in prices and quantities of malt bought for the ports of London, Plymouth and Dover by the Commissioners of the Victualling Board

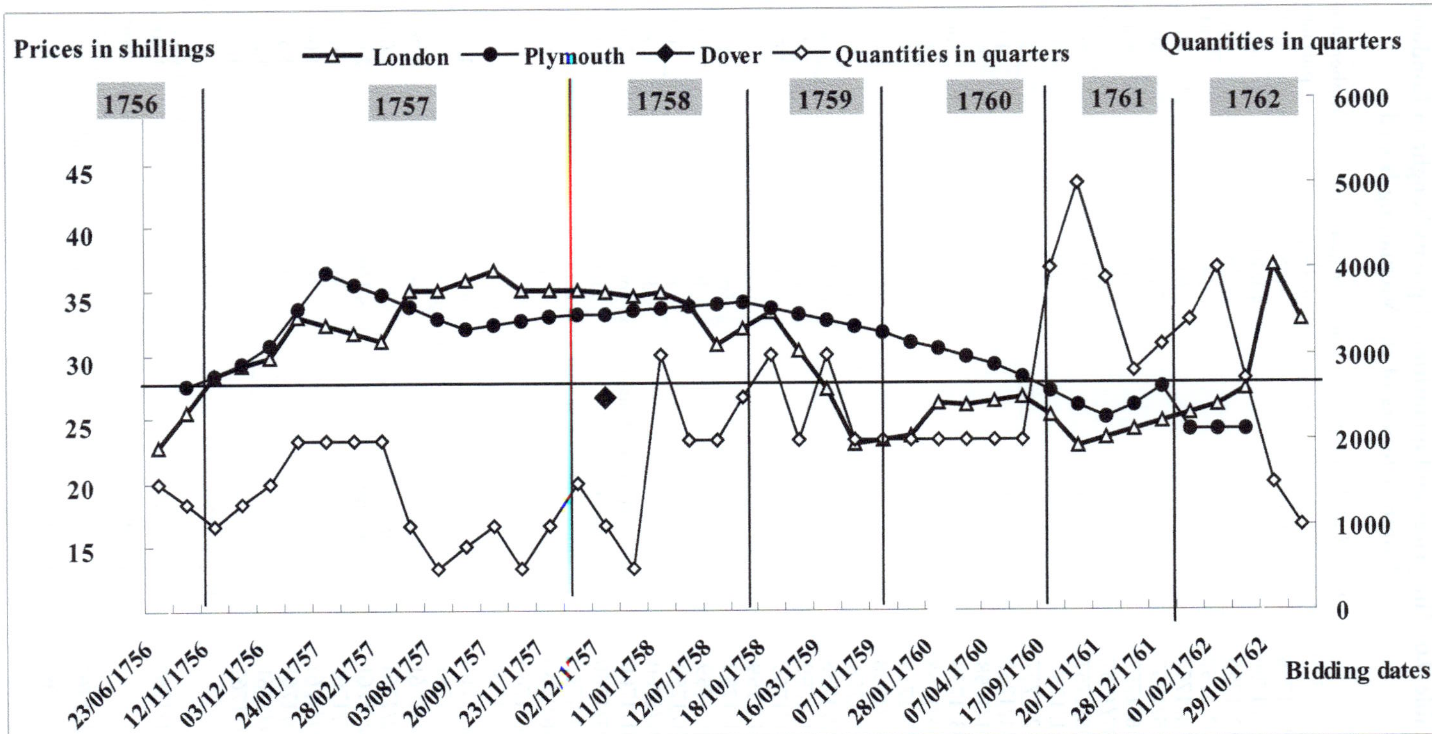

make one ton of sea beer.[8] Twenty-eight of the aforementioned contracts were issued for provisions for the port of London, thirteen for Plymouth, and one contract was for Dover, dated 2 December 1757 and requiring 500 quarters to be delivered within eight days.[9]

The Plymouth contracts stated that the contractors had a fortnight in which to load their malt into merchant vessels *en route* for Plymouth, whilst the London contracts did not specify dates or gaps between deliveries. The only stipulation was the clause '*on demand*', which signified that the contractor had to be ready at a moment's notice to deliver the number of quarters of malt requested – this could be a fraction of, or the totality of, the sum he had been contracted to supply. The amount of time at his disposal in which to deliver the produce depended on the acuteness of the need. Time intervals between contracts also fluctuated, because the Commissioners would only organise a tender for bids when the preceding quantity was on the way to being delivered.

Generally speaking, the call to tender bids for malt in Plymouth took place on the same day as those for wheat, and the time limits for delivery and loading were identical. This was a financial measure, designed to reduce costs by transporting both products in a single convoy.

Furthermore, contracts for malt of the highest quality ('*without any mixture of brown or pale malt*') specified the year in which germination had to take place. Most often, '*new malt*' was provided, that is to say, malt that had been germinated between 1 October and 30 April of the current year. From time to time, however, the Commissioners chose to buy malt that had been prepared the year before, called '*old malt*', and sometimes purchases combined the two.

Figure 9.7 shows the price curves for malt destined for London and Plymouth. An examination of these curves reveals the same tendencies as for the other products highlighted in this chapter. In fact, prices began to climb in 1756 and reached their highest point in 1757, then they began to decrease in 1758 and were at their lowest in 1759, 1760 and 1761, before sharply increasing again in 1762. There are, however, some specific points to be noticed: malt prices never fluctuated by 50 per cent as was the case for wheat, oats and groats, but only by a maximum of 35 per cent. Malt price reductions also occurred later and in a more progressive manner, with the decline beginning in 1757 but never truly confirmed until 1759 (and not 1758). There was also a very marked elasticity between prices and quantity. The Commissioners opted, in 1756–7, to increase the number of invitations to tender bids in order to limit the quantities required from each contract (see

8 PRO, ADM 30/44, p. 258.

9 PRO, ADM 111/46, 2 December 1757.

Table 9.7 The prices and quantities of malt bought for the ports of London, Plymouth and Dover by the Commissioners of the Victualling Board

	London		Plymouth	Dover
Bidding date	Price per quarter (in shillings)	Quantity in quarters	Price per quarter (in shillings)	Price per quarter (in shillings)
23/06/1756	22/5	1,500		
06/10/1756			27/7	
12/11/1756	28/4	1,000		
17/11/1756			29/3	
03/12/1756	29/10	1,500		
05/01/1757	33	2,000		
24/01/1757			36/5	
09/02/1757			Offer at price of last or next contract	
28/02/1757	31	2,000		
04/05/1757	35	1,000		
03/08/1757	35	1,000		
24/08/1757			31/11	
26/09/1757	36/6	1,000		
02/11/1757	35	500		
23/11/1757			32/11	
02/12/1757				29/10
09/01/1758	34/6	500		
11/01/1758	34/10	3,000		
17/03/1758	33/11	2,000		
12/07/1758	30/9	2,000		
13/10/1758			34	
18/10/1758	33/3	2,000		
12/01/1759	?	?		
16/03/1759	27/2	3,000		
24/10/1759	23	2,000		
07/11/1759			31/8	
07/12/1759	23/8	2,000		
28/01/1760	26/2	2,000		
21/03/1760	26	2,000		
07/04/1760			29/2	
21/05/1760	26/7	2,000		
17/09/1760	25/3	4,000		
11/02/1761	22/10	5,000		
20/11/1761			25	
09/12/1761	24	2,800		
28/12/1761			27/4	
27/01/1762			24	
01/02/1762	25/11	4,000		
26/03/1762			24	
29/10/1762	36/10	1,500		
01/12/1762	32/8	1,000		

Figure 9.7). On the other hand, as from 1758, and even more markedly in 1760, 1761 and 1762, the Commissioners no longer hesitated to buy up to 5,000 quarters at a time, which represented more than twice the amount contracted to a single contractor in the invitations to tender bids of 1756 and 1757, in which most contracts were for only 500 or 1,000 quarters (see Figure 9.7). Consequently, the number of contracts decreased in spite of the increase in quantities delivered.

Peas

Between 1 April 1756 and 31 December 1762 there were no fewer than fifty-four invitations to tender bids in order to stock the London warehouses with peas. No contract concerning the victualling of another port was to be found in the minutes of the Central Board, which would imply that once again these provisions were bought and delivered to London and then periodically sent on to Dover, Portsmouth and Plymouth (see Chapter 2). These ports would sometimes complete their provisions by buying a few items locally, under the jurisdiction of the Agent in charge of each port.

Peas were also measured in terms of quarters, and were delivered by the bushel of 64 pounds to Redhouse and Hartshorn, according to proportions drawn up in each contract.[10]

The contracts were drawn up according to a fairly consistent rhythm in terms of regularity of delivery – every month or so – even though this was not as regimented as was the case for wheat. This time scale was not even modified by the sudden rise in purchases that came into effect from 1759, in which every contract was simply allocated a greater number of quarters (see Figure 9.8). Price fluctuations followed the same tendencies as those already described for wheat.

In conclusion, it is striking to notice the progressive and appreciable reduction in the prices of the foodstuffs throughout the 1758–61 period, even though, at the same time, the volumes purchased increased substantially. This surprising development puts into question a number of fundamental assumptions which will be addressed in the general conclusion of this book. However, one can already postulate that these lowered prices helped make possible the marked increase in Navy numbers which in turn was at the heart of British naval and military success from 1759. A final issue, of more limited consequence, and which will be dealt with in the second part of this chapter, is to consider the extent to which the bigger contracts (issued from 1758–9) were responsible for the ousting of smaller suppliers who were unable to provide the huge quantities now requested.

10 PRO, ADM 30/44, p. 258.

Figure 9.8 Fluctuations in prices and quantities of peas bought in London by the Commissioners of the Victualling Board

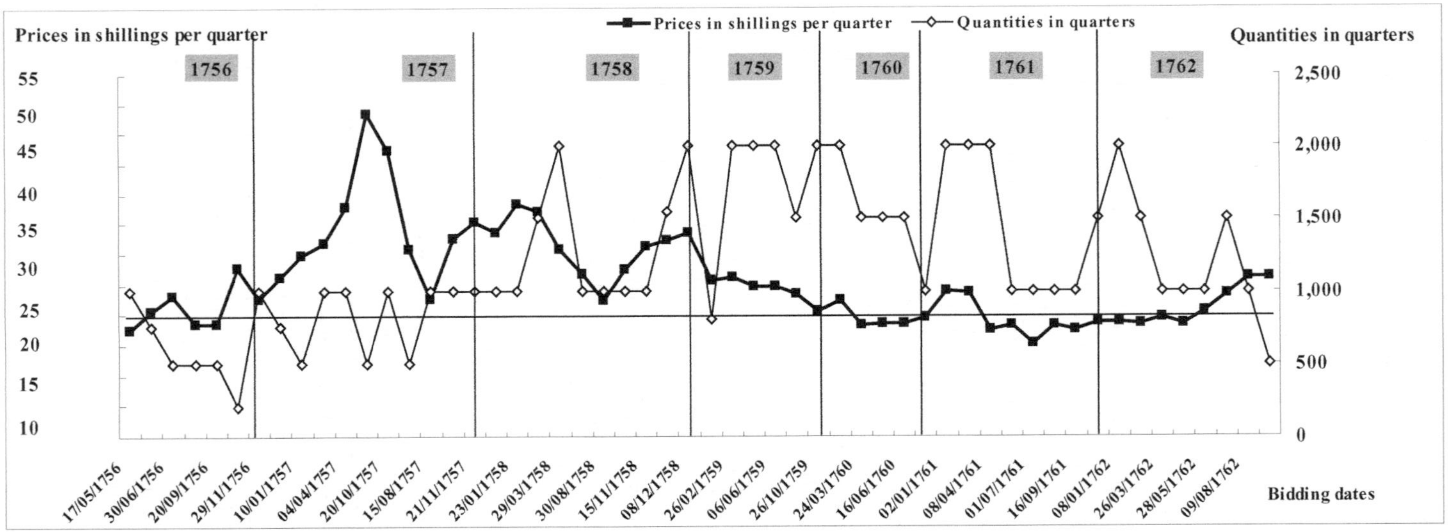

Table 9.8 The prices and quantities of peas bought in London by the Commissioners of the Victualling Board

Bidding date	Period of delivery by weekly instalments	Price per quarter (in shillings)	Number of quarters purchased
17/06/1756	1 month	22	1,000
18/06/1756	Price too high: no contract made		
30/06/1756	1 month	27/5	500
16/08/1756	"	22/9	500
20/09/1756	"	22/10	500
20/10/1756	15 days	32	200
20/11/1756	1 month	26/9	1,000
05/01/1757	Price too high: no contract made		
10/01/1757	1 month	34	500
02/03/1757	"	36	1,000
04/04/1757	"	42	1,000
11/05/1757	3 weeks	57/6	500
20/06/1757	1 month	51/6	1,000
13/07/1757	"	35	500
15/08/1757	"	26/11	1,000
07/10/1757	"	36/8	1,000
21/11/1757	"	39/5	1,000
21/12/1757	"	37/10	1,000
23/01/1758	"	42/8	1,000
13/02/1758	3 weeks	41/4	1,500
20/03/1758	6 weeks	35	2,000
28/06/1758	1 month	30/10	1,000
30/08/1758	6 weeks	26/8	1,000
06/10/1758	1 month	31/9	1,000
15/11/1758	"	35/2	1,000
06/12/1758	Price too high: no contract made		
08/12/1758	1 month	37/9	2,000
31/01/1759	"	29/9	800
26/02/1759	6 weeks	30/5	2,000
06/04/1759	"	28/9	2,000
06/06/1759	1 month	28/9	2,000
13/07/1759	"	27/7	1,500
26/10/1759	"	24/8	2,000
21/01/1760	6 weeks	29/9	2,000
21/01/1760	1 month	30/5	1,500
21/01/1760	"	38/9	1,500
21/01/1760	6 weeks	27/7	1,500
21/01/1760	1 month	24/8	1,000
02/01/1761	14 days	28/1	2,000
18/02/1761	6 weeks	27/9	2,000
08/04/1761	1 month	21/10	2,000
27/05/1761	"	22/6	1,000
01/07/1761	"	19/5	1,000
14/08/1761	"	22/5	1,000
16/09/1761	"	21/10	1,000
28/10/1761	"	21/10	1,500

08/01/1762	5 weeks	22/11	2,000
01/03/1762	1 month	22/8	1,500
26/03/1762	"	23/8	1,000
30/04/1762	"	22/10	1,000
28/05/1762	"	24/10	1,000
07/06/1762	"	27/8	1,500
09/08/1762	"	30/5	1,000
27/08/1762	15 days	30/4	500

The Body of Contractors and What It Reveals

The Commissioners of the Victualling Board in London issued a total of 320 contracts for cereals and pulses[11] between 1 April 1756 and 31 December 1762.[12] Of the eighty-eight individuals who took part in this invitation to tender bids, only fifty (56.82 per cent) obtained one or more contracts (Table 9.10).[13]

Table 9.9 The number of invitations to tender bids per sector of activity

Wheat for London	88	Groats	45
Wheat for Plymouth	16	Flour	46
Peas	54	Malt for London	28
Oats	29	Malt for Plymouth	14
Total = 320			

The extreme concentration of this market is most noticeable when the number of contracts won by the seven most important contractors is taken into consideration (cf. Table 9.2). Messrs Bidwell (38), Cowley (17), Farrar (75), Fossett (41), Millward (27), Parker (18) and Smith (22) shared 238 of the possible 320 contracts, which represents 74.37 per cent of the total market.

Each of these important suppliers will be presented, as well as the degree of market atomicity. Finally, the conditions that engendered true competition will be analysed.

11 The biscuit sector (cf. Figure 9.4) was not included.

12 An inventory of all contracts issued in the final quarter of 1760 could not be drawn up due to gaps in the ADM 111 archival series.

13 Individuals who were also members of five associations of traders are included in this total.

Table 9.10 The number of individuals who obtained contracts

Surname –	First name	Wheat for London	Wheat for Plymouth	Oats	Peas	Groats	Flour	Malt for London	Malt for Plymouth	Total
Bennes	John	-	-	-	-	-	-	-	1	1
Bells		-	-	-	-	-	-	1	-	1
Bidwell	Thomas	1	1	5	15	-	-	11	5	38
Bisson	Daniel	-	-	-	-	-	3	-	-	3
Bradley	Thomas	-	-	1	2	2	-	-	-	5
Brown	Peter	-	1	-	-	-	-	-	-	1
Burnett	John	-	-	1	-	-	-	-	-	1
Burnett	Robert	-	-	-	1		-	1	-	2
Church	Thomas	-	-	-	-	4	-	-	-	4
Constable	Abraham	-	-	1	-	-	-	-	-	1
Cowley	Richard	11	5	1	-	-	-	-	-	17
Curtis	J.	-	-	-	6		1	-	-	7
Easl	William	-	1	-	-	-	-	-	-	1
Everett	William	-	-	-	-	-	-	9	-	9
Farrar	Thomas	32	11	4	11	-	8	6	3	75
Fossett	Edward	20	-	4	8	-	6	1	2	41
Greens		-	1	-	-	-	-	-	-	1
Hankin	Edward	2	-	-	-	-	-	-	-	2
Heron	William	1	3	-	-	-	-	-	3	7
Hilbert	John	-	-	-	-	-	1	-	-	1
Hill	Edmund	-	-	-		-	7	-	-	7
Hollingsworth	Samuel	1	-	1	1	-	-	2	-	5
Keene	Thomas	2	-	-	-	-	5	-	-	7
Lefevre	John	5	-	-	-	-	3	-	-	8
Merret	Alexandre	-	-	-	1	-	-	-	-	1
Millward	J., T. and W.	5	-	3	7	4	8	-	-	27
Morgan	John	-	-	-	-	4	-	-	-	4
Nash	Edward	-	-	-	-	5	6	-	-	11
Nelson	Georges	-	-	-	1	-	-	2	-	3
Newman	Abraham	-	-	1	-	-	-	4	-	5
Oldham	John	-	-	3	1	3	-	-	-	7
Parker	William	-	-	2	-	16	-	-	-	18
Peacock	John	-	-	-	-	1	-	-	-	1
Ratcliffe	R.	-	-	-	1	-	-	-	-	1
Ruse	John	-	-	1	2	-	-	-	-	3
Sally	S. and I.	1	-	-	-	-	-	-	-	1
Scofield	Thomas	7	-	-	-	-	-	-	-	7
Shalett	Edward	4	-	-	-	-	-	-	-	4
Shaw	Samuel	-	-	-	2	-	-	-	-	2
Smith	Charles	5	-	-	-	-	17	-	-	22
Smith	John	-	-	-	-	-	9	-	-	9
Symons	Thomas	-	-	-	-	-	-	-	1	1

Tackle	John	-	-	-	-	1	-	-	-	1
Trottman	John	1	-	-	1	-	-	1	-	3
Williams	Henry	-	-	1	-	-	-	-	-	1
Wilson	Robert	4	-	-	1	-	-	-	-	5
Woodard	John	-	-	-	2	-	-	-	-	2
Woods	William	-	1	-	-	-	-	-	1	2
Messrs Burnett and Brooks		-	-	2	-	-	-	-	-	2
Messrs Burnett and Hollingsworth		-	-	-	-	-	-	1	-	1
Messrs Curtis and Rondeau		1	-	-	-	-	-	-	-	1
		103	**24**	**31**	**63**	**40**	**74**	**39**	**16**	**390**[1]

Merchants acting independently 48
Associations 3
Business entities 50

1 The 320 contracts (cf. Table 9.1) were broken down into 390 contracts, since some contracts had been divided up between several merchants.

The main suppliers

Two sub-groups can be identified within the group of the seven most important contractors: three of them were specialist suppliers whilst the other four dealt in a variety of products.

Within the specialised group was Richard Cowley who dealt almost exclusively with wheat – indeed, sixteen of his seventeen contracts were for wheat, while only one was for oats. Research carried out in the Corporation of London Record Office revealed two items of information concerning him. The CFI/539 register reveals that he was taken on as an apprentice weaver in April 1733, the date on which his name appeared in the Bureau of London Corporations register. Then, the London *Trade Directory* of 1760, also housed in the Record Office, mentions the names of 'Cowley and Burnett, Distillers, of Great Drivery Lane'. This is almost certainly a reference to John and/or Robert Burnett who participated in three tenders for bids for wheat and who was/were offered a few contracts for oats and malt. John Burnett also entered into association with William Brook, while Robert Burnett entered into association with Samuel Hollingsworth, and the two groups respectively were awarded two contracts for oats and one for malt; Samuel Hollingsworth independently obtained three contracts for victualling of the Port of London in oats, wheat and peas, and two contracts for malt (see Table 9.10). This example clearly illustrates the links that existed between merchants, links that would have been lost in the examination of Victualling Board documents alone, in which no mention is made of a partnership with Richard Cowley.

The other member of the specialised group was Charles Smith, who only placed bids for wheat and for flour. Although Smith was not a particularly successful merchant in the wheat sector, and only obtained five contracts out of forty-eight proposals, he was, on the other hand, the main supplier of flour, with a total of seventeen contracts which represented 22.97 per cent of the market. He outstripped a certain John Smith[14] who obtained nine contracts and with whom he was in direct competition on a number of occasions. These two men succeeded in obtaining more than one-third of the total market share.

William Parker specialised in the victualling of groats, a market that he succeeded in dominating since he held 40 per cent of all contracts (sixteen out of forty contracts). He also participated in three tenders for the oats market and obtained two contracts in this sector.

The final four merchants were involved in more diversified areas of business. Economically speaking they carried much more weight, having obtained 181 of the 320 available contracts (56.56 per cent), as opposed to a total of 17.81 per cent for the first three contractors profiled above. Table 9.11 summarises their areas of business interest and their relative importance in the victualling market.

Two of these four contractors were the leading suppliers of six out of the eight types of provision under consideration. Thomas Farrar was the most important supplier of wheat, both for London and for Plymouth; Thomas Bidwell was the main supplier of oats, peas and malt for London and Plymouth; finally, the groats and malt markets were dominated by William Parker and Charles Smith.

Farrar and Bidwell were more successful (in terms of percentages) than Edward Fossett, the Millwards (John, Thomas and William) and other suppliers. Out of 174 and 93 invitations to tender bids respectively, Farrar obtained a total of 75 contracts, representing 43.10 per cent of the market, and Bidwell obtained 38, representing 40.86 per cent of the market. Edward Fossett and the Millwards participated in 141 and 140 invitations to tender bids, but only obtained 29.07 per cent and 19.28 per cent of the market respectively (see Table 9.11). Even though Edward Fossett's sectors of activity were more spread out, he was nonetheless the second-most important supplier in terms of total numbers of contracts obtained (41), in which he surpassed Thomas Bidwell (38 contracts) and the Millwards (27 contracts).

An examination of the Corporation of London Record Office archives is, in fact, rather disappointing. Mention is made in May 1727 of a certain Thomas Farrar who was a cutler's apprentice,[15] and of a William Millward, who was

14 John Smith lived in Great St Helens and held the status of esquire (cf. Corporation of London Records Office, Trade Directory, 1760).

15 Corporation of London Records Office, C.F.1/473.

Table 9.11 A breakdown of contracts obtained by the four most important suppliers

	Thomas Bidwell				Thomas Farrar			
Products	Participation	Contracts awarded	Rating	Market share	Participation	Contracts awarded	Rating	Market share
Wheat for London	9	1		0.97% (1/103)	62	32	No. 1	31.07% (32/103)
Wheat for Plymouth	2	1	No. 4	4.17% (1/24)	18	11	No. 1	45.83% (11/24)
Oats	11	5	No. 1	16.13% (5/31)	15	4	No. 2 with Fossett	12.90% (4/31)
Groats	-	-		0% (0/40)	8	0		0% (0/40)
Flour	1	-		1.35% (1/74)	19	8	No. 3	10.81% (8/74)
Malt for London	23	11	No. 1	28.21% (11/39)	20	6	No. 3	15.38% (6/39)
Malt for Plymouth	12	5	No. 1	31.25% (5/16)	9	3	No. 2	18.75% (3/16)
Peas	35	15	No. 1	23.81% (15/63)	23	11	No. 2	17.46% (11/63)
	93	**38**			**174**	**75**		
	40.86%				**43.10%**			

	Edward Fossett				John, Thomas and William Millward			
Products	Participation	Contracts awarded	Rating	Market share	Participation	Contracts awarded	Rating	Market share
Wheat for London	63	20	No. 2	19.42% (20/103)	42	5		4.85% (5/103)
Wheat for Plymouth	7	0		0% (0/24)	3	0		0% (0/24)
Oats	11	4	No. 2 with Farrar	12.90% (4/31)	18	3	No. 3	9.68% (3/31)
Groats	-	-		0% (0/40)	24	4		10% (4/40)
Flour	12	6		8.11% (6/74)	22	8	No. 3	10.81% (8/74)
Malt for London	9	1		2.56% (1/39)	-	-		0% (0/74)
Malt for Plymouth	3	2	No. 4	12.5% (2/16)	-	-		0% (0/16)
Peas	36	8	No. 3	12.70% (8/63)	31	7	No. 4	11.11% (7/63)
	141	**41**			**140**	**27**		
	29.07%				**19.28%**			

an upholder's apprentice,[16] but both these jobs were so far removed from the victualling sector that it is extremely unlikely that either of these men was one of our merchants. However, the 1760 *Trade Directory* does mention a Thomas Farrar, *corn-factor*, at 27 Prescot Street, Goodman's Fields.[17]

Study of the bids made and contracts obtained by these four merchants shows that Thomas Bidwell, the most important supplier of oats, did not obtain any further contracts for supplies of this cereal after 1757; Thomas Farrar participated in fewer and fewer invitations to tender bids, as the years went by, in the peas, malt and groats sectors; Edward Fossett became more and more important as time went by in the malt and flour sectors; the Millwards, an ever-present force in the oats, peas and groats sectors in 1756–8, then redirected their efforts to the wheat and flour markets in the ensuing years. The wheat market was dominated by Farrar in 1759–60 and then by Fossett in 1761–2. A total of thirty-four contracts for provisions of wheat to London were obtained by these two merchants in 1759, 1760, 1761 and 1762.

It is also interesting to note that Thomas Farrar participated in four tenders for bids for cheese and butter[18] (see Chapter 10), and Edward Fossett participated in a tender for bids for the provision of meat to the port of Portsmouth.

The different market sectors and their respective levels of competition

Over the eight types of product concerned, Table 9.12 illustrates the fact that about twenty merchants were competing in all sectors of the market. For example, the market for wheat destined for the victualling yard at Plymouth had twenty contracts; the market for peas had twenty-three contracts; oats had twenty contracts; groats had twenty-two contracts; flour had twenty-two contracts; and malt had nineteen contracts. However, wheat destined for London had a total of thirty-five merchants competing for contracts, and the malt market destined for Plymouth only had thirteen suppliers.

The number of merchants who obtained contracts, expressed in terms of percentages, reveals the degree to which the markets were concentrated. The market for oats was the least concentrated, with 75 per cent of all competitors obtaining one or more contracts.

Table 9.13 shows the market share obtained in each sector by the four main merchants. The oats sector seems to have been the least concentrated, followed by the flour sector, in which the four main suppliers only succeeded in obtaining approximately 50 per cent of the contracts: 51.61 per cent for oats, 56.75 per cent for flour. Conversely, in the case of the wheat and malt sectors destined for Plymouth and in the malt sector destined for London,

16 Corporation of London Records Office, C.F.1/507.

17 Corporation of London Records Office, Trade Directory, 1760.

18 On 15 September 1756, 16 August 1758, 15 August 1759 and 8 September 1760.

Table 9.12 Competitors and suppliers to each sector of the market

	Wheat for London	Wheat for Plymouth	Peas
Competitors	**35** (34 individual merchants, 1 association)	**20** individual merchants	**23** (21 individual merchants, 2 associations)
Competitors sharing the market	**15** (42.86%)	**8** (40%)	**17** (73.91%)
	Oats	**Groats**	**Flour**
Competitors	20 (18 individual traders, 2 associations)	22 individual merchants	22 individual merchants
Competitors sharing the market	**15** (75%)	**11** (50%)	**13** (59.09%)
	Malt	**Malt for Plymouth**	
Competitors	**19** (17 individual merchants, 2 associations)	**13** individual merchants	
Competitors sharing the market	**11** (57.89%)	7 (53.85%)	

Table 9.13 Market shares obtained in each sector by the four main suppliers

Wheat for Plymouth			**Malt for Plymouth**		
Total number of contracts	**24**		***Total number of contracts***	**16**	
Farrar	11	45.83%	Bidwell	5	31.25%
Cowley	5	66.66%	Farrar	3	50.00%
Heron	3	79.16%	Heron	3	68.75%
Bidwell	1	83.33%	Fossett	2	81.25%
Malt			**Wheat**		
Total number of contracts	**39**		***Total number of contracts***	**103**	
Bidwell	10	25.64%	Farrar	32	31.07%
Everett	9	48.72%	Fossett	20	50.49%
Farrar	6	64.10%	Cowley	11	61.17%
Newman	4	74.36%	Scofield	7	67.96%
Groats			**Peas**		
Total number of contracts	**40**		***Total number of contracts***	**63**	
Parker	16	40.00%	Bidwell	15	23.80%
Church	4	50.00%	Farrar	11	41.27%
Morgan	4	60.00%	Fossett	8	53.97%
Oldham	3	67.50%	Millward	7	65.07%
Flour			**Oats**		
Total number of contracts	**74**		***Total number of contracts***	**31**	
Smith Ch.	17	22.97%	Bidwell	5	16.12%
Smith John	9	35.14%	Farrar	4	29.03%
Farrar	8	45.95%	Fossett	4	41.93%
Millward	8	56.75%	Millward	3	51.61%

there was a monopolistic tendency in which three-quarters of all contracts were won by a small minority of merchants. Farrar alone obtained 45.83 per cent of all wheat contracts for London and William Parker obtained 40 per cent of all contracts for groats. The four main suppliers held 65.07 per cent of all contracts for peas, 67.96 per cent for wheat for London and 67.50 per cent for groats.

Real or imaginary competition?

Under what conditions was the competition for market shares really carried out? Any complicity that may have existed between traders is much harder to uncover in this sector than it was in the meat sector. The cereals and pulses sector seems to have had only five merchant associations – an insufficient quantity by which to recreate the networks of suppliers – unlike the networks that were discovered in the meat sector (cf. Chapter 8). Instead, a different approach has been taken in researching this sector: an examination of the effect that the number of competitors had on prices, and the effect that the number of competitors had on the quantities requested by the Victualling Board. Figure 9.9 shows the number of merchants who participated in each invitation to tender bids for the victualling of the port of London yards in wheat, since it is for this product that the largest number of contracts was issued.

The movement of prices in this case, when compared to average annual prices, reveals that there was genuine competition taking place. The high prices of 1756, 1757 and 1758 coincide with greater numbers of competitors: 4.93 in 1756, 5.87 in 1757 and 5.53 in 1758. On the other hand, lowered prices in 1759, 1760 and 1761 coincide with a reduction in the number of competitors. This trend was particularly marked in 1761, a year which had the lowest prices and the lowest number of competitors (2.69 – see Figure 9.1).

An examination of available correspondence reveals only one allusion to a cartel – this was in a letter concerning the granting of a contract for wheat: '*the Board thinking the above prices too high and that there is a combination ...*'.[19]

Sometimes spontaneous offers would be made, which also corroborates the notion that genuine competition existed. From time to time, a merchant who had not participated in the invitation to tender bids would spontaneously offer his services for the provision of a certain quantity of foodstuffs. The Commissioners would sometimes accept these offers, but only on the condition that the price per quarter matched that of the last or subsequent contract: '*on the terms of the present or the next contract at the Board's option*'.[20]

19 PRO, ADM 111/47, 26 June 1758.

20 PRO, ADM 111/48, 4 December 1758.

Figure 9.9 Fluctuations in the number of merchants competing for contracts in the wheat sector

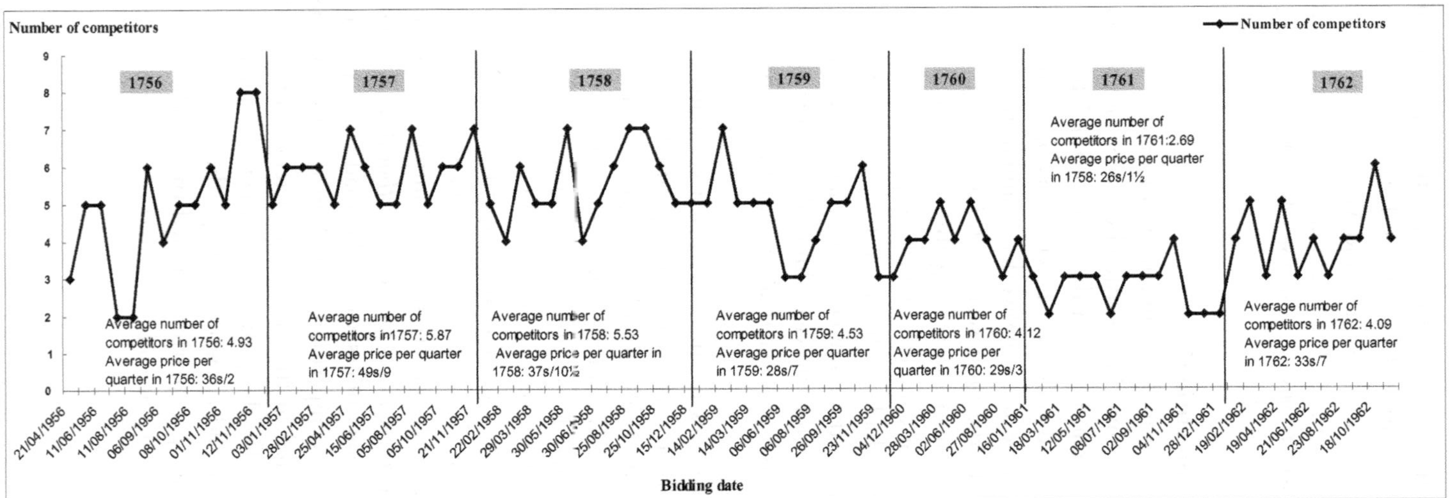

This ensured that the Commissioners could not be accused of collusion and permitted the existence of both a competitive system (invitations to tender bids) and a spontaneous system.

Did an increase in required quantities lead to a decrease in the number of merchants capable of meeting the demand, and thus to the creation of cartels, especially in the last years of the war? An examination of the wheat market (Figure 9.9) shows that this was not the case, quite simply because a group of merchants could each simultaneously take responsibility for a part of the required foodstuff, providing, of course, that they agreed on the lowest price. This was also quite often the case for the flour market, in which two suppliers would share the market. During the record demands for flour on 5 and 11 January 1761, five and seven contractors respectively delivered the 3,000 quarters requested.[21] As far as malt is concerned, the record acquisition of 5,000 quarters on 11 February 1761 involved four contractors who all tendered bids and offered the same tariff per quarter.[22] Only two of the four merchants, however, had the means to meet a demand for 5,000 quarters. This explains why, despite the sharp increase in demand, no tension was felt over prices (cf. Figure 9.7). The same reasoning can be applied to wheat, flour and peas (Figures 9.1, 9.3 and 9.8). The only exception was for groats, which may have led to the special procedure introduced for this product at the beginning of 1762: apart from their regularly scheduled invitations to tender bids, the Commissioners contacted a Mr Edward Nash on 15 January 1762 in order to arrange for a delivery of 50 or 60 quarters of groats per week. The Commissioners proposed Nash a price that was one shilling less per quarter than the contracts they had established that same day during the invitation to tender bids, with the proviso that either they or Nash could terminate the agreement with two weeks' notice.[23] This very specific agreement lasted throughout the war, with the prices changing periodically.[24] On 19 April, for example, Nash's price for a quarter of groats was 48 shillings (based on that day's invitation to tender bids minus one shilling). This was four shillings higher than the price per quarter on 15 January.[25]

The increase in quantities purchased by the Victualling Board in the final years of the Seven Years War did not lead to a reduction in conditions of genuine competitiveness. On the contrary, apart from the year 1762 in which prices rose due to the bad harvest, the Commissioners' system of invitations

21 PRO, ADM 111/51, 5 and 11 January 1761.

22 PRO, ADM 111/51, 11 February 1761.

23 PRO, ADM 111/53, 15 January 1762.

24 PRO, ADM 111/53, 19 April, 14 June 1762; ADM 111/54, 20 October, 1 November 1762.

25 PRO, ADM 111/53, 19 April 1762.

to tender bids resulted in a net decrease in prices. If there were any tensions at all, these were usually due to delivery dates imposed by the Commissioners, which could not always be met by the suppliers. In these cases, the Commissioners had the right to charge for the product at the previously fixed price, or to ask for a new price based on their current analysis of the market – and once again, these agents of the Crown would find themselves in the position of arbiters.[26]

26 PRO, ADM 111/45, 31 December 1756; ADM 111/53, 15 February 1762; ADM 111/54, 24 September 1762.

10

The Remaining Sectors: Beverages, Butter and Cheese, Salt, Olive Oil and Raisins

Having examined homogeneous sectors such as meat, cereals and pulses, this final chapter will concentrate on a group of sectors which, although rather more disparate, was of no less importance. In fact, 104 merchants – as individuals or in associations – were involved in supplying the Navy with provisions of butter, beer, cheese, olive oil, raisins, salt, spirits and wine.

Supplying the Navy with Beverages

Beer

Very few contracts were drawn up for the provision of the various Navy yards with beer, and there is a logical explanation for this. During the War of the Austrian Succession and into the 1750s, the Navy made a special effort to construct and maintain breweries in its main ports (see Chapters 1 and 2). During the Seven Years War, according to the historian P. Mathias,[1] it was therefore able to provide for three-quarters of its needs in beer. Mathias quotes a contemporary as noting that this allowed the Victualling Board

> To provide most of the beer required by the Navy at a far less expense as well as in better quality than what is furnished in contracts; the Crown will also thereby be secured from the ill effects of combinations too frequently found among contractors where they have it in their power to make their own terms ... the contract brewers never brewing but as the beer has been demanded so that in cases of emergency the Service has been at all times subject to delay and disappointments and especially in their brewing beer in the summer season.[2]

Only two invitations to tender bids were required by the London yards to ensure that their stocks of beer were complete, so efficient was the Victualling Board's brewery in Hartshorne near Tower Hill. The first invitation to tender bids in January 1757 resulted in the purchase of 1,200 tuns which had

1 Mathias, *The Brewing Industry*.

2 Mathias, *The Brewing Industry*, 202.

to be delivered within a fortnight.[3] The second invitation to tender bids was in September 1757 and resulted in the purchase of 150 tuns.[4] The twelve suppliers who participated in these invitations to tender bids proposed the same price on both occasions, which would suggest that there was a certain complicity between the merchants, which was undoubtedly damaging to the Crown (see Table 10.1).

Table 10.1 Merchants competing for two purchases of sea beer destined to complete the port of London's stocks

Invitation to tender bids of 3 January 1757 (ADM 111/45)				
Merchants' names	**Quantities offered (in tuns)**	**Delivery time**	**Price per tun**	
Lawrence Ambrose	120	14 days	£3/12/0	None of these merchants reduced their prices at the second meeting with the Commissioners
Peter Champion	140	"	"	
Thomas Curtis	60	"	"	
William Collison,	220	"	"	
Matthew Dove esquire	140	"	"	
David Frinder	50	"	"	
Henry Goodwin	40	"	"	
Messrs R. Hare and Salmon	140	"	"	
Samuel Hawley	50	"	"	
John Lefevre	80	"	"	
Messrs Seranche and North	100	"	"	
Henry Willoughby esquire	60	"	"	
	1,200 tuns bought by the Commissioners of the Victualling Board			

Invitation to tender bids of 14 September 1757 (ADM 111/46)			
Merchants' names	**Quantity required 150 tuns**	**Delivery time set to 10 days**	**Price per tun**
Peter Champion			£3/12/0
William Collison			"
Richard Hare			"
Ja. Seranche			"
Henry Willoughby esquire (could not supply the required quantity)			
The Commissioners bought from the four suppliers who were able to ensure the stipulated 37.5 tuns			

If the Victualling Board's brewery in South Down near Plymouth was also efficient enough to fulfil the port's requirements in beer, this was not the case for Portsmouth. Indeed, it was the only port that came under the jurisdiction of the Victualling Board to need regular victualling in beer. Every February, or the beginning of March at the latest, the Commissioners would organise an

3 PRO, ADM 111/45, 3 January 1757.

4 PRO, ADM 111/45, 14 September 1757.

invitation to tender bids and would choose one contractor who would then be responsible for whatever quantities of beer were asked of him for the duration of one year, the contract having been dated as starting on 13 March. He also had to be in a position to provide any one of a variety of beers: sea beer or petty warrant beer, or harbour beer,[5] which was part of the rations when on dry land and had a lower level of alcohol (see Chapter 3).

A dozen merchants competed for this contract throughout the war, and it was won on four occasions (1756, 1757, 1760 and 1761) by Sir Thomas Ridge, an MP in the House of Commons and a successful merchant, for he was considered to be the Navy's main supplier of beer at Portsmouth.[6] The contract was also awarded on two occasions to a member of his family (who was nonetheless in competition with him), George Ridge Esquire.

Figure 10.1 illustrates that there was a logical, parallel price evolution for both sea beer and land beer. By February 1759, after having experienced a slight increase in prices in 1758, the price of beer had dropped by one-third of the February 1757 price.

Figure 10.1 Fluctuations in the price of sea beer and 'land' (petty warrant) beer bought by the Commissioners of the Victualling Board for the port of Portsmouth

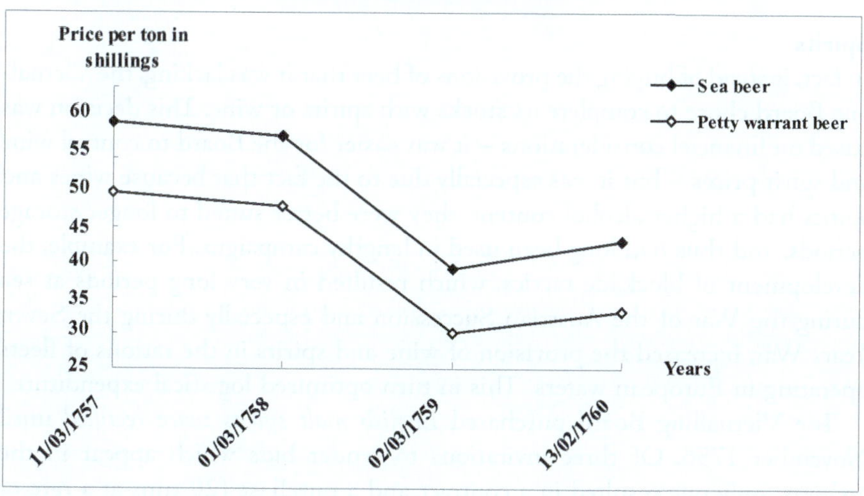

A letter written by Sir Thomas Ridge on 13 November 1761 indicates that despite being a full eight months into his one-year contract, not a single

[5] Mathias, *The Brewing Industry*, 199.

[6] Baugh, *British Naval Administration*, 399.

order had been placed.[7] It is perhaps not surprising that in the following year (March 1762) not a single merchant was willing to participate in the invitation to tender bids. This marked the end of Portsmouth's regular victualling in beer.[8]

Table 10.2 The prices of beer bought by the Commissioners of the Victualling Board for Portsmouth

Bidding date	Price per tun of sea beer (shillings/pence)	Price per tun of petty warrant beer (shillings/pence)
11/03/1757	58	48
01/03/1758	?	?
02/03/1759	39/11	29/11
13/02/1760	43	33

It would be interesting to know how the Victualling Board managed to satisfy its beer requirements without having recourse to outside suppliers, given that, at best, it produced three-quarters of its needs. The survey that follows of the wine and spirits market is rich in information and will throw a light on this question.

Spirits

In fact, instead of buying the provisions of beer that it was lacking, the Victualling Board chose to complete its stocks with spirits or wine. This decision was based on financial considerations – it was easier for the Board to control wine and spirit prices – but it was especially due to the fact that because wines and spirits had a higher alcohol content, they were better suited to longer storage periods, and thus had long been used in lengthy campaigns. For example, the development of blockade tactics, which resulted in very long periods at sea during the War of the Austrian Succession and especially during the Seven Years War, increased the provision of wine and spirits in the rations of fleets operating in European waters. This in turn optimised logistical expenditure.

The Victualling Board purchased *English malt spirits twice rectified* until November 1756. Of three invitations to tender bids which appear in the archives, only one resulted in a contract and a purchase (20 tuns at a rate of £37/19s a tun);[9] the tariffs proposed at the two other invitations to tender

7 PRO, ADM 111/52, 13 November 1761.

8 PRO, ADM 111/53, 10 March 1762.

9 PRO, ADM 111/44, 11 August 1756.

bids were considered as being much too high.[10] Table 10.3 also clearly shows the presence of a number of suppliers of meat.

Table 10.3 Merchants competing to supply the Commissioners in malt spirits and brandy

A – Malt spirits: 8 merchants	
James Adams*	John Lefevre*
Mark Bell	Johnson Pasco*
Joseph Bishop	
Isaac Olier*	
John Cook*	George Shaw

* A merchant who had also participated in one or more bids for meat contracts

B – Brandy: 6 merchants	
John Biggin	William Grant
Francis Fanning	Anthony Malcom
John Fierney	Francis Smith

Of the nine contracts:

John Biggin obtained	4[1]
Francis Smith obtained	4[2]
Francis Fanning obtained	1[3]

1 Cf.: 15 December 1758 (see 15 January 1759), 2 January, 27 February 1761, 13 January 1762.
2 Cf.: 2 December 1760, 13 January, 14 April, 16 June 1762.
3 Cf.: 2 November 1761.

However, growing evidence that appeared to link the consumption of malt spirits with an increase in the number of cases of scurvy brought provisions of malt spirits to an abrupt end.[11] In fact, once supplies of sea beer had run out, the men's rations included malt spirits instead, and it was noticed that scurvy was making a come-back. Consequently, on 27 November 1756, the Lords of the Admiralty forbad all further deliveries of malt spirits and ordered the Victualling Board's agents to sell all remaining quantities at public auctions.[12]

Malt spirits were gradually replaced by French brandy or, on occasion, Flemish brandy.[13] These provisions were acquired not by invitations to tender bids, but by spontaneous offers from individual merchants.[14] (This

10 PRO, ADM 111/44, 30 July and 8 November 1756.

11 PRO, ADM 1/924. Boscawen to the Admiralty, dated 12 February 1757, cited in Gradish, *The Manning of the British Navy*, 155.

12 PRO, ADM 111/45, 29 November 1756.

13 PRO, ADM 111/51, 27 February 1761.

14 PRO, ADM 111/48, 29 January 1759; ADM 111/52, 3 August 1761.

does not include provisions of brandy and rum destined for Louisburg and Belle-Ile.)[15]

Figure 10.2 Fluctuations in the price of brandy and quantities bought by the Commissioners of the Victualling Board

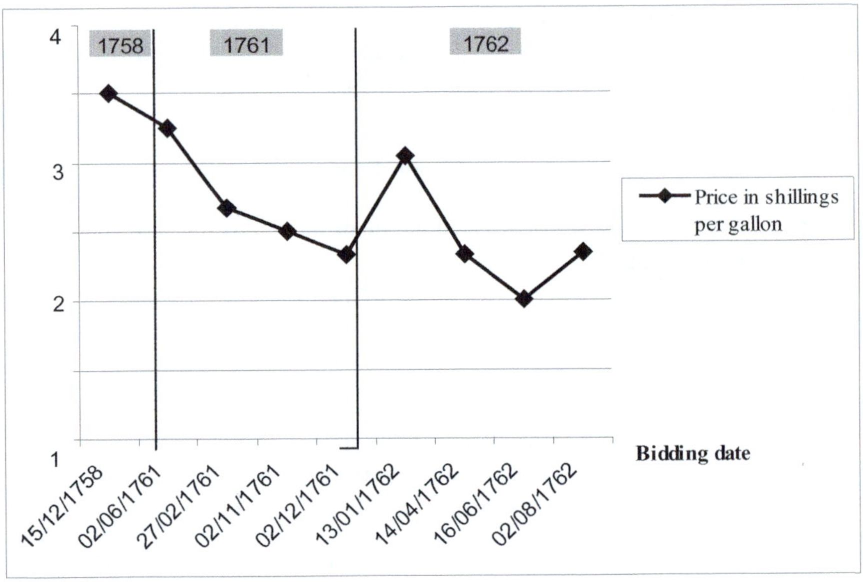

Table 10.4 The prices of brandy bought by the Commissioners of the Victualling Board

Bidding date	Price in shillings/pence per gallon
15/12/1758	3/6
02/01/1761	3/3
27/02/1761	2/8
02/11/1761	2/6
02/12/1761	2/4
13/01/1762	3/0½
14/04/1762	2/4
16/06/1762	2/0
02/08/1762	2/4

15 PRO, ADM 111/52, 2 December 1761.

Wine

In terms of volume, deliveries of wine to ports under the jurisdiction of the Victualling Board increased dramatically during the course of the war. Five hundred pipes were bought per year in 1756, 1757 and 1758; the quantity rose in 1759 and 1760 to a combined amount of 1,750 pipes,[16] and reached an all-time high in 1761 with the acquisition of 3,000 pipes of wine (see Figure 10.3). When this development is combined with that of spirits during the last two years of the war, it is clear why the Victualling Board was able to dispense with the services of private brewers in completing its stocks of beer.

With the exception of Vidonia wine from the Canary Islands, which was only bought on one occasion in November 1758,[17] two appellations of wine shared the market: a white wine from Malaga and a red wine from Oporto. Each type of wine had its own particular suppliers. Alexander Pepper had an almost total monopoly on Vidonia wine, as did Manuel Vieira for wine from Oporto. A number of merchants, however, competed for contracts for Malaga wine. Amongst them were Abraham Kirkpatrick, Christopher Roberts, Messrs Clark and Robinson, Anthony Bacon, Maud, Baynes and John Fierney.

Manuel Vieira's position was even stronger during the last two years of the war. This was due to the rising number of complaints received regarding Malaga wine, which led to a decision, taken in February 1761, that the wine did not keep well in hot climatic conditions and was responsible for an outbreak of intestinal and gastric illness amongst the crews.[18] Until then, wine purchases had been divided more or less equally between Oporto and Malaga wine.[19] Now, only Portuguese wine was to be bought, which made Mr Vieira the Victualling Board's sole supplier of wine for the main capital yards.[20]

The suppliers of wine were expected to guarantee that it would keep for a period of six months following delivery. The international scope of wine victualling, just as for brandy, meant that the Commissioners had to gain the Lords of the Admiralty's approval before embarking on any transaction. Furthermore, the limited competition that prevailed in this sector led to the Commissioners organising fewer and fewer invitations to tender bids, and, instead, to a tendency to respond to spontaneous propositions which were regularly forthcoming.

16 Taking into consideration the fact that the 1760 archives are missing figures for the last quarter.

17 PRO, ADM 111/48, 3 November 1758.

18 PRO, ADM 111/51, 18 and 27 February, 9 March 1761.

19 PRO, ADM 111/51, 27 February 1761.

20 PRO, ADM 111/52, 14 October and 17 November 1761; ADM 111/54, 21 July, 4 August, 13 and 24 September 1762.

Figure 10.3 Fluctuations in the price of wine and quantities bought by the Commissioners of the Victualling Board

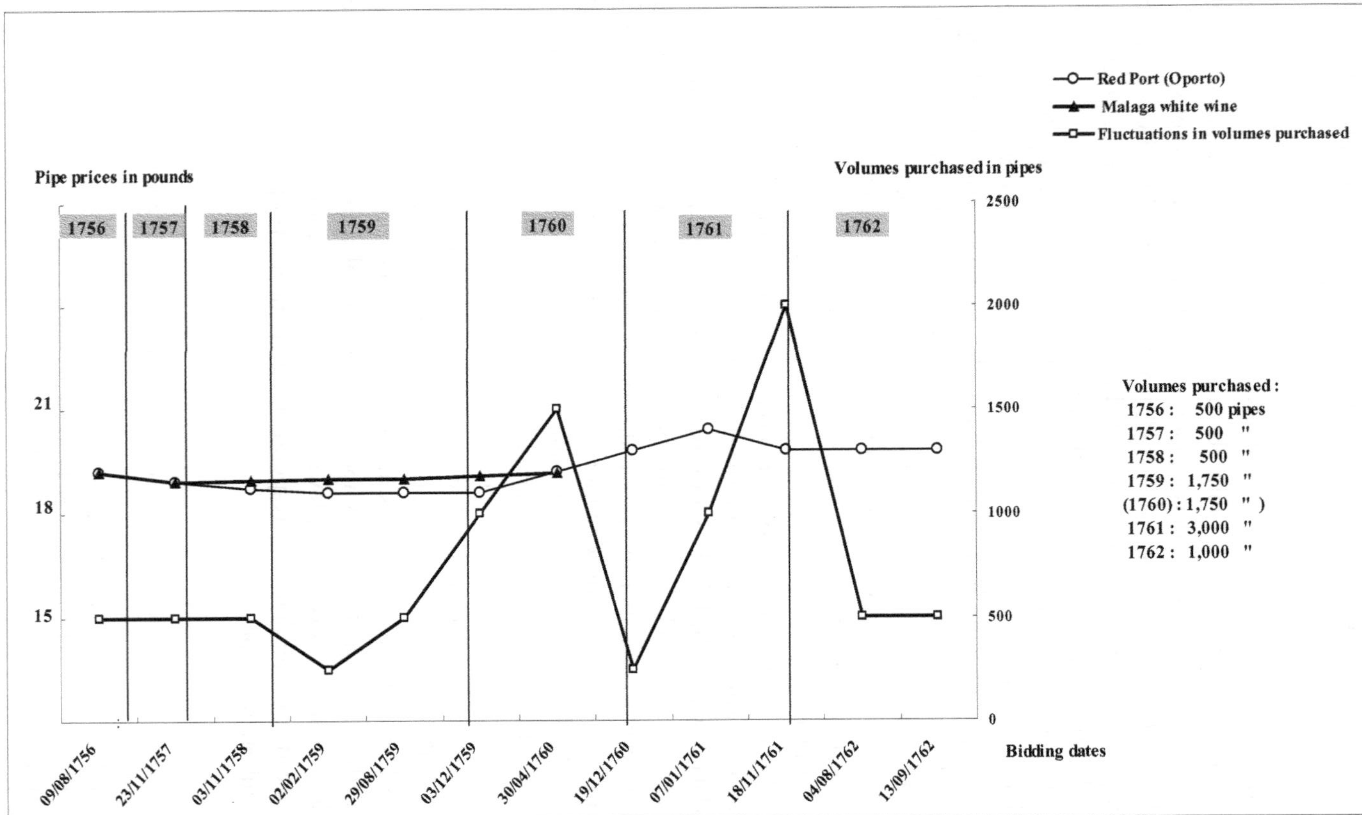

Table 10.5 The prices of wine bought by the Commissioners of the Victualling Board

Bidding date	Porto red wine Price of pipe in £ sterling	Malaga white wine Price of pipe in £ sterling	Canaries wine Price of pipe in £ sterling
09/08/1756	18/10	18/10	
13/11/1757		18/5	
03/11/1758			21
02/02/1759	18		
29/08/1759	18		
03/12/1759	18		
30/04/1760	18/10	18/19	
19/12/1760		?	
07/01/1761	19/10		
18/11/1761	19		
04/08/1762	19		
13/09/1762	19		

Because of the fact that taxes were imposed on imported products and especially on alcoholic beverages, imported wines were exclusively delivered to Guernsey where they were exempt from import duty due to the island's privileged status until December 1759. Wine on Guernsey was stocked under the responsibility of a Victualling Board Agent, Mr Saumerez,[21] and would be sent, as dictated by need, directly on board ships anchored at Portsmouth or Plymouth, or at any other meeting point, without having transited on English soil.

This method bordered on the absurd. Victualling Board Agents in the ports quickly began to voice their dissatisfaction, claiming that the procedure was too long, especially when sailing against the prevailing winds, too risky and, above all, too costly. Supported by the Commissioners, they formally pointed out to the Lords of the Admiralty, in a letter dated 19 October 1759, that it would be much more advantageous to deliver wine directly to Portsmouth or Plymouth.[22] Their request was heeded, and the Lords succeeded in gaining reluctant permission from the Treasury to unload wine in the ports of Portsmouth and Plymouth without paying an import duty on it, but only in cases where it was destined for consumption on warships. It had to be '*kept under the locks of the officers of the custom and issued to His Majesty's Ships when the service should require, duty free*'.[23] Consequently, on 3 December 1759, for

21 PRO, ADM 111/44, 11 August 1756.

22 PRO, ADM 110/19, 19 October 1759.

23 PRO, ADM 110/20, 21 April 1760.

the first time ever, the Commissioners contracted for a supply of wine which would be delivered directly to the ports of Portsmouth and Plymouth.[24] The Victualling Board would have to wait until the end of 1761 to be able to proceed in a similar way as regards the introduction of spirits on to English soil.[25]

The only point to be made as far as tariffs are concerned is that there was a slight increase in prices (of just under 6 per cent) during the last two years of the war (see Figure 10.3), which illustrates the extent to which prices were successfully controlled.

Annual Contracts for Butter, Cheese and Salt

Butter and cheese

One single contract offered on an annual basis to a merchant assured the Navy's provisions of butter and cheese. The contract was allocated in mid-September during the first two years of the war, and was then put forward to mid-August in order to give the supplier enough time to acquire the provisions and have them sent to London. Although the contract was allocated in August, it actually came into effect immediately following Michaelmas Day, on 30 September, and ended on 29 September the following year.

The contract for butter and cheese was rather unusual not only because it was an annual contract, but also because it did not specify any fixed quantities. In fact, the contractor had to be in a position to supply the Victualling Board's warehouses in London[26] with any quantity at any given time. It was then up to the agents of the Victualling Board to dispatch the produce from the London base to whichever yard was in need.

Factors such as the length of the contract, the unpredictability of the contractor's obligations, and the sheer quantities in question (1,365,000 lbs of butter and 2,730,000 lbs of cheese, based on annual rations of 39 lbs of cheese per man and 19½ lbs of butter per man for a force of 70,000 men[27]) meant that only the most important merchants could take part in the invitation to tender bids. To undertake the delivery of unknown quantities at a previously agreed rate for a period of a year represented a risk that only the major merchants were in a position to take, and clarifies why only three or four merchants felt confident enough to be present at the invitation to tender bids. The cheese and butter sector, which has already been the object

24 PRO, ADM 111/49, 3 December 1759.

25 PRO, ADM 111/52, 2 December 1761; ADM 111/53, 14 April, 16 and 21 June 1762.

26 PRO, ADM 110/18, 7 September 1757.

27 This figure was often larger. Cf. Gradish, *The Manning of the British Navy*, 42.

of careful study by G.E. Fussell,[28] C. Goodman[29] and R.B. Westerfield,[30] and thus is well documented, was also, geographically speaking, extremely concentrated. In fact, these three or four major suppliers were all based in London and, according to a contemporary, formed a sort of club:

> The most considerable cheesemongers in London have formed themselves into a club. They are owners of about sixteen ships. They have factors in Cheshire who buy up the cheese for them and lodge it in their warehouse at Chester. At their weekly meetings they settle what quantities each shall have bought up to market. By this means the market is fed in such a manner, that they command whatever price they please.[31]

The fleet of sixteen ships that they controlled ferried butter and cheese from York, Hull, Scarborough, Stockton, Newcastle, Suffolk, Chester and Liverpool to London. In 1729 alone, London dealt with 21,000,000 lbs of cheese transiting through the port.[32]

Only two companies shared in the victualling of cheese and butter during the Seven Years War: Messrs Robert Barnevelt and Thayer Townshend were granted the contract in the first five invitations to tender bids (1756–60) and Messrs Joseph Smith and James Copeland obtained the final two contracts (1761 and 1762). Robert Barnevelt had already taken responsibility for most of the Navy's provisions of cheese and butter during the war of 1739–48.[33] Seven other merchants (including Thomas Farrar, the Victualling Board's main supplier of cereals) participated from time to time in the invitations to tender bids, but with no success whatsoever (see Table 10.6).

The end of Barnevelt and Townshend's reign was provoked by a serious incident in 1761 that set the two merchants against the Commissioners. Following a complaint made by the captain of the *Royal Sovereign* in Portsmouth on 26 January 1761, the Commissioners proceeded to open – in front of Robert Barnevelt – a number of barrels of cheese that had recently arrived.[34] The cheese was not considered to be of the desired quality, and so an inquiry was opened.[35] In the space of two months, every other barrel of

28 G.E. Fussell, 'The London cheesemongers of the eighteenth century', *Economic History* (Jan. 1928), 394–8.

29 G.E. Fussell and C. Goodman, 'The eighteenth century traffic in milk products', *Economic History* (Feb. 1937), 380–7.

30 R.B. Westerfield, *The Middleman in English Business* (New Haven, Yale University Press, 1915).

31 Wimpey, cited in Westerfield, *The Middleman*, 208.

32 W. Maitland, *The History and Survey of London* (London, 1756), II, 756–9.

33 Baugh, *British Naval Administration*, 412–16.

34 PRO, ADM 111/51, 26 January 1761.

35 PRO, ADM 111/51, 28 January 1761.

Table 10.6 Merchants competing for the victualling of the Royal Navy in butter and cheese

Merchants who took part in the invitation to tender bids	
***Merchants acting individually* 6**	
Thomas Farrar	Edmund Pridle
Humphrey Mortimore	John More
Thomas Price	James Smith
***Associations* 4**	
Messrs Robert Barnevelt and Thayer Townshend	
Messrs Tho. Lewis and Tho. Fossick	
Messrs John More and Joseph Smith	
Messrs Joseph Smith and James Copeland	
Breakdown of the awarded contracts	
1756: Messrs Robert Barnevelt and Thayer Townshend	
1757: Messrs Robert Barnevelt and Thayer Townshend	
1758: Messrs Robert Barnevelt and Thayer Townshend	
1759: Messrs Robert Barnevelt and Thayer Townshend	
1760: Messrs Robert Barnevelt and Thayer Townshend	
1761: Messrs Joseph Smith and James Copeland	
1762: Messrs Joseph Smith and James Copeland	

cheese was opened in the yards of London, Chatham, Dover, Portsmouth and Plymouth. The results were conclusive: the barrels of inferior-quality cheese had been placed in such a way amongst the other barrels that the contractors' argument that this was due to an accident was simply not supported. The Commissioners, who were sure beyond reasonable doubt that this was an act of '*premeditated design*', decided to sue Barnevelt and Townshend for breach of contract.[36] The two contractors declared that they had unknowingly been the victims of one of their agents, and pleaded with the Commissioners not to pursue the case, promising to cover all costs.[37] Unfortunately, there are no records stating whether or not the case did go to court, but it is known that they did not participate in the last two invitations to tender bids. This opened up a no doubt extremely lucrative market for Messrs Smith and Copeland.[38]

Another example demonstrates the seriousness with which the Commissioners took their task: their role in modifying the different types of cheese with which they were victualled. The cheeses that were transported to London came from the four regions clearly stated in the invitation to tender bids: Suffolk, Gloucestershire, Warwickshire and Cheshire. Having observed, in 1758, a number of complaints stating that Suffolk cheese preserved less

36 PRO, ADM 111/51, 27 April 1761.

37 PRO, ADM 111/51, 27 April 1761.

38 PRO, ADM 111/52, 24 August 1761; ADM 111/54, 25 August 1762.

well than the others, the Commissioners set up an investigation. They were soon able to exonerate the contractor of any suspicion of fraud, and it became clear that the problem lay in the nature of the cheese itself. The investigation revealed that cheese from this region of the country remained soft for quite a long time after being made, which rendered it unsuitable for the Navy's very particular needs. The Commissioners therefore requested that the cheese be stocked in an open, well-ventilated place and wiped down on a daily basis to keep it dry – instructions that were almost impossible to execute in a period of great demand.[39]

Due to these almost impossible working conditions, the Commissioners of the Victualling Board requested permission from the Lords of the Admiralty to no longer buy Suffolk cheese, and simply to continue purchasing the Cheshire, Warwickshire and Gloucestershire cheeses. Despite the fact that the latter cheeses cost more than Suffolk cheese (see Figure 10.4), the Lords readily agreed,[40] and decided that from then on, one-third of all cheese requirements should be met by Cheshire cheeses. The contracts now stipulated that all cheeses had to have been made at least four months before delivery, and six months before delivery in the case of Cheshire cheeses. The Victualling Board provided the barrels in which to put the cheese or butter, but the supplier had to cover packaging costs,[41] and guarantee that the product would keep for six months following delivery.

Figure 10.4 Fluctuations in the price of butter and cheese bought by the Royal Navy

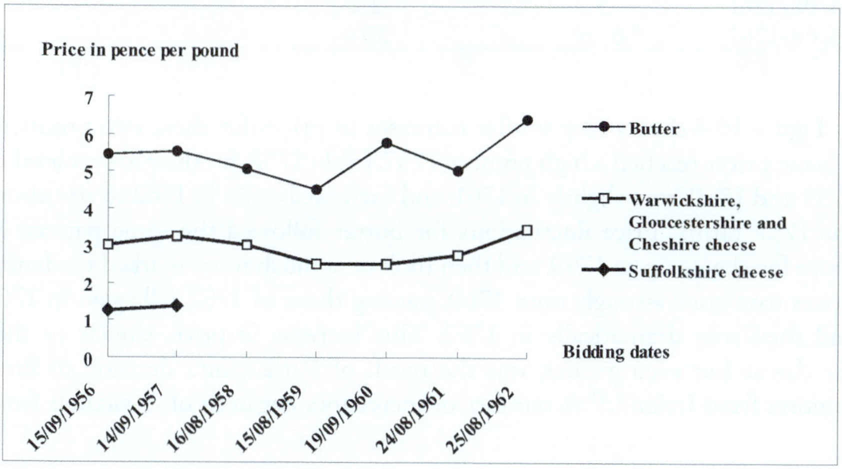

39 PRO, ADM 110/19, 26 and 27 July 1758.

40 PRO, ADM 111/47, 28 July 1758.

41 PRO, ADM 111/47, 16 August 1758; ADM 111/49, 15 August 1759; ADM 111/50, 13 August 1760.

The situation concerning the provision of butter was a little more complex since the contractors were strictly forbidden from using butter of Irish origin. This ban was not simply a result of the embargo that was periodically applied to Irish imports, since the Navy had had access to the Irish market as from 1758, but rather resulted from a decision made by the Commissioners of the Victualling Board alone. They were of the opinion that imports from Ireland would impede the smooth running of the delivery system due to the fact that, according to the statutes, imported Irish butter had to be delivered directly to the Victualling Board's warehouses. The Commissioners opted to dispense with this source of butter, at the risk of inflated prices, and to continue benefiting from the more expensive but regular delivery system that had already been established in which the suppliers were given a mere five days in which to supply the requested amounts of butter and cheese demanded of them.[42]

Table 10.7 The prices of butter and cheese bought by the Royal Navy

Bidding date	Price per pound of butter (in pence)	Price per pound of Warwickshire, Gloucester and Cheshire cheese (in pence)	Price per pound of Suffolk cheese (in pence)
15/09/1756	5 7/16	3	1¼
14/09/1757	5½	3¼	1⅜
16/08/1758	5 1/6	3	-
15/08/1759	4½	2½	-
19/09/1760	5¾	2½	-
24/08/1761	5	2¾	-
25/08/1762	6 3/83	3 7/16	-

Figure 10.4 shows very similar increases in prices for these two products. Cheese prices reached a high point in 1757, fell in 1758, hit their lowest level in 1759 and 1760, rose slightly in 1761, and increased again in 1762 to rise above the 1757 tariffs. Price fluctuations for butter followed the same pattern as those for cheese up to 1759, and then took on a much more marked tendency: prices rose quite strongly until 1760, passing those of 1757, fell again in 1761 and then rose dramatically in 1762. This increase in price, similar to that for cheese but even greater, was the result of Parliament's decision to limit imports from Ireland.[43] A number of merchants began to offer victuals from

42 PRO, ADM 111/53, 2 April 1762 reveals the Victualling Board's refusal to authorise the import of Irish butter. Only one exception was made, for a delivery that was not covered by a contract, in January 1760. Cf. PRO, ADM 111/50, 18 January 1760.

43 PRO, ADM 111/53, 2 April 1762.

Ireland in 1762, but since only the Royal Navy now had the right to obtain provisions from Ireland, as described earlier in this chapter, these offers were declined by the Commissioners of the Victualling Board.[44]

Salt

Contracts for salt were also awarded annually. The main difference between these contracts and those for butter and cheese is that the contracts for salt stipulated the quantities to be delivered to the four main victualling bases at London, Portsmouth, Plymouth and Dover. These quantities were measured in weys or bushels depending on the type of salt being furnished. On the same day, the Commissioners would issue two contracts per port corresponding to the two main types of salt: white salt and bay or brine salt. Three-quarters of the Navy's needs in salt were provided by white salt (see Table 10.8).

The origin of white salt is clearly indicated in the archival records: it either came from Newcastle (Newcastle salt) or from Lymington, to the west of Portsmouth (Lymington salt). Indeed, Lymington's geographical position meant that Lymington salt was the more economical option for the ports of Dover, Plymouth and, of course, Portsmouth. Newcastle salt and Lymington salt only entered into competition as far as the London market was concerned. In fact, London was supplied with Newcastle salt twice, in 1757 and 1759, and with Lymington salt four times, in 1758, 1760, 1761 and 1762.

Bay and brine salt were the objects of separate contracts, and were also competing products. In 1758, 7,990 bushels of brine salt and 2,080 bushels of Cagliari salt (from a region in southern Sardinia) were bought. In 1759, 3,000 bushels of English bay salt were bought. However, throughout all the other years of war, only brine salt was ever bought.

Brine salt was prepared according to the method invented in 1741 by Thomas Lowndes. A document found in the ADM series reveals that in the 1750s the Commissioners carried out a variety of experiments in order to test that Lowndes's method really did produce salt best suited to their needs.[45] The results obviously proved to be conclusive, for the salt produced by Lowndes's method was judged to be '*equal or to excelle French bay salt for domestick uses and for curing beef or pork for the Navy*'. However, its use in the preservation of fish was questioned, following a number of contradictory reports. As a result of these experiments, a total of 41,250 bushels containing 56 lbs each of brine salt were purchased by the Commissioners in the years 1754–8.[46]

44 PRO, ADM 111/54, 25 August 1762.

45 PRO, ADM 110/19, 29 January 1759.

46 PRO, ADM 110/19, 29 January 1759.

Table 10.8 A breakdown by category of annual sales of salt

	1757	1758	1759	1760	1761	1762
White salt						
from Newcastle	350 weys	-	900 weys	-	-	700 weys
from Lymington	430 weys	1,060 weys	324 weys	740 weys	940 weys	530 weys
Total in pounds	780 x 40 = 31,200 bushels	1,060 x 40 = 42,400 bushels	1,224 x 40 = 48,960 bushels	740 x 40 = 29,600 bushels	940 x 40 = 37,600 bushels	1230 x 40 = 49,200 bushels
	31,200 x 56 = 1,747,200 lbs	42,400 x 56 = 2,374,400 lbs	48,960 x 56 = 2,741,760 lbs	29,600 x 56 = 1,657,600 lbs	37,600 x 56 = 2,105,600 lbs	49,200 x 56 = 2,755,200 lbs
Brine salt and bay salt						
Brine salt	9,660 bl x 56 = 540,960 lbs	7,990 bl x 56 = 447,440 lbs	9,857 bl x 56 = 551,992 lbs	10,150 x 56 = 568,400 lbs	12,568 x 56 = 703,808 lbs	13,200 x 56 = 739,200 lbs
Bay salt	-	2 080 bl x 84 = 174,720 lbs	3 000 bl x 84 = 252,000 lbs	-	-	-
Total in pounds	540,960 lbs	622,160 lbs	803,992 lbs	568,400 lbs	703,808 lbs	739,200 lbs
Total in pounds	2,288,160 lbs	2,996,560 lbs	3,545,752 lbs	2,226,000 lbs	2,809,408 lbs	3,494,400 lbs
% of white salt	76.36%	79.24%	77.33%	74.47%	74.95%	78.84%
% of brine salt and of bay salt	23.64%	20.76%	22.67%	25.53%	25.05%	21.16%

Table 10.9 Merchants competing for the victualling of the Royal Navy in white salt

Merchants who participated in the invitation to tender bids 8		
Robert Ballard		
Messrs Biggin and Ward		
James Dixon		
John Le Keux		
John Stuart		
James White		
Thomas Williams		
Merchants who participated in the six auctions 4 (1757, 1758, 1759, 1760, 1761 and 1762)		
James Dixon		
John Stuart		
James White		
Thomas Williams		
Merchants who obtained one or more contracts 5		
James Dixon		
2 contracts:	350 weys of Newcastle salt for London	(14 February 1757)
	900 weys of Newcastle salt for London	(21 March 1759)
John Le Keux		
1 contract:	30 weys of Lymington salt for Dover	(28 January 1761)
John Stuart		
7 contracts:	90 weys of Lymington salt for Portsmouth	(21 March 1759)
	44 weys of Lymington salt for Dover	(21 March 1759)
	500 weys of Lymington salt for London	(19 March 1760)
	53 weys of Lymington salt for Dover	(19 March 1760)
	230 weys of Lymington salt for Portsmouth	(14 April 1762)
	200 weys of Lymington salt for Plymouth	(14 April 1762)
	100 weys of Lymington salt for Dover	(14 April 1762)
James White		
1 contract:	112 weys of Lymington salt for Plymouth	(19 March 1760)
Thomas Williams		
12 contracts:	220 weys of Lymington salt for Portsmouth	(14 February 1757)
	160 weys of Lymington salt for Plymouth	(14 February 1757)
	50 weys of Lymington salt for Dover	(14 February 1757)
	870 weys of Lymington salt for London	(3 March 1758)
	50 weys of Lymington salt for Portsmouth	(3 March 1758)
	140 weys of Lymington salt for Plymouth	(3 March 1758)
	190 weys of Lymington salt for Plymouth	(21 March 1759)
	75 weys of Lymington salt for Portsmouth	(19 March 1760)
	650 weys of Lymington salt for London	(28 January 1761)
	90 weys of Lymington salt for Portsmouth	(28 January 1761)
	170 weys of Lymington salt for Plymouth	(28 January 1761)
	700 weys of Lymington salt for London	(14 April 1762)
Total number of contracts: 23		

Figure 10.5 Fluctuations in the price of white salt destined for the ports of London, Portsmouth, Plymouth and Dover

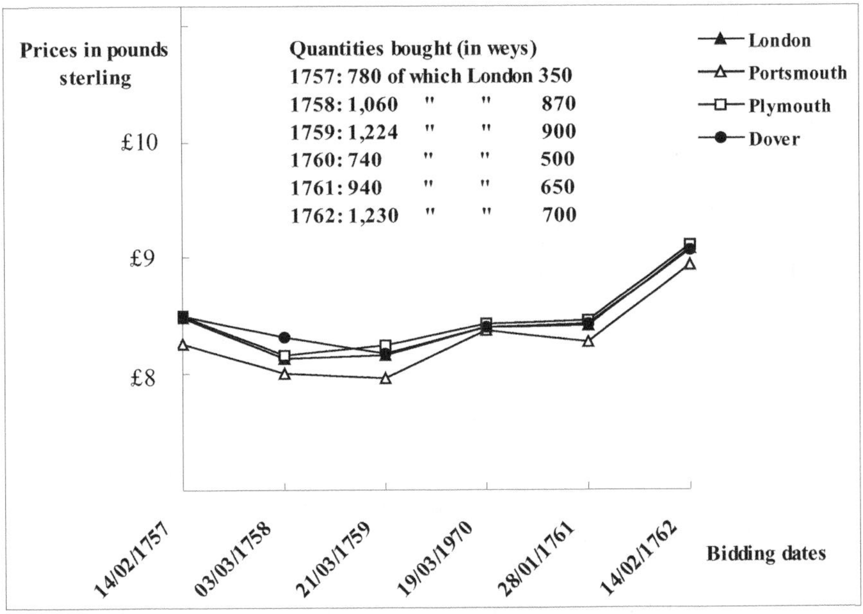

Figure 10.6 Fluctuations in the price of brine salt and bay salt

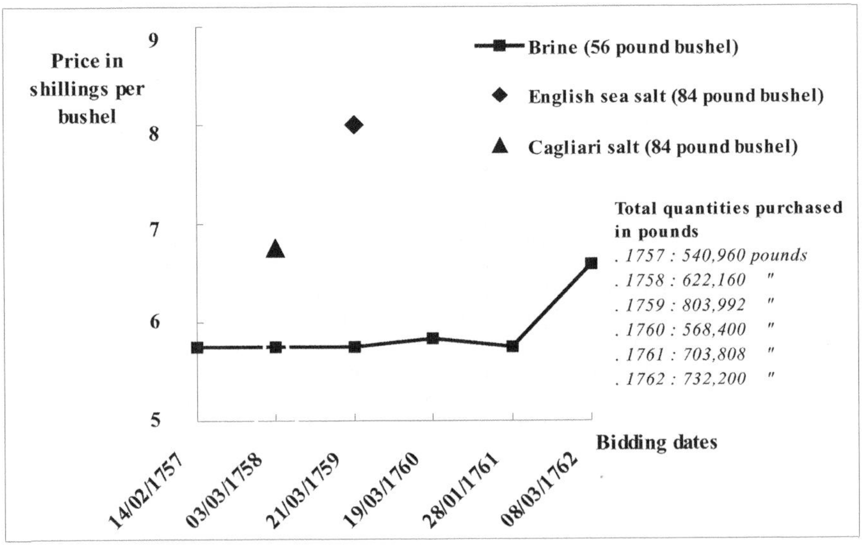

Whether the product was white salt, brine salt or bay salt, a strict time-table regulated the rhythm of supply.[47] Half the salt had to be delivered before 30 September, which marked the beginning of the slaughter season for cattle and pigs, and the remaining half before Christmas.

Similarly to the cheese and butter market, the salt market lay within the control of a small group of merchants. Eight traders would put in an appearance from time to time at the invitation to tender bids for white salt (see Table 10.9), but only four of them took part in the six tenders, and shared twenty-two of the twenty-three available contracts between them. They were James Dixon, John Stuart, James White and Thomas Williams. Mr Williams alone was awarded twelve contracts, including some of the most important ones, such as the London contracts in 1758 and 1761. Amongst the unlucky traders was John Biggin, who had opted to associate himself with a certain Mr Ward on this occasion. They had participated by mail in the invitation to tender bids in 1762.

The brine salt sector was practically a monopoly.[48] Isaac Wood, the supplier of brine salt, did not encounter any competition until 1761 and consequently was awarded all contracts from 1754 to 1760. A trader called John Mackay made his appearance in 1761 and controlled the market in 1761 and 1762.[49]

In terms of prices, it is interesting to note that white salt followed the same movement as the rates for butter – albeit less markedly. The rise in prices in 1760 and the high point that was reached in 1762 are comparable to those for butter. These tariffs were no doubt the result of the embargo on cured meat from Ireland. A similar tendency was found in the tariffs for brine salt, which rocketed in 1762 after years at the same level (see Figure 10.6). Nonetheless, white salt destined for Portsmouth was obtained at a constantly lower price than for the other three ports (see Figure 10.5).

47 Except for the year 1762, in which the first half of the supply had to be delivered before 30 June, and the second half before 30 September. Cf. PRO, ADM 111/53, 8 March 1762.

48 Provisions of bay salt were not included because this product was victualled in insignificant quantities. Also, the three traders who offered English bay salt were the same men who dealt in white salt, so their names already appear in the list of merchants.

49 PRO, ADM 111/51, 28 January 1761; ADM 111/53, 8 March 1762.

Table 10.10 The prices of white salt for London, Portsmouth, Plymouth and Dover

Bidding date	Price of white salt for London in £ per wey	Price of white salt for Portsmouth in £ per wey	Price of white salt for Plymouth in £ per wey	Price of white salt for Dover in £ per wey
14/02/1757	8/9/4	8/2/10	8/9/9	8/9/9
03/03/1758	7/18/9	7/15/0	7/19/10	-
21/03/1759	7/19/10	7/14/0	8/2/4	8/0/0
10/03/1760	8/7/0	8/6/0	8/8/0	8/7/0
28/01/1761	8/7/6	8/3/4	8/8/9	8/8/0
14/02/1762	9/7/6	9/3/0	9/8/0	9/7/0

Table 10.11 The prices of brine salt and bay salt purchased for the Royal Navy

Bidding date	Price of brine salt in shillings per bushel	Price of English bay salt in shillings per bushel	Price of Cagliari salt in shillings per bushel
14/12/1757	5/9	-	-
03/03/1758	5/9	-	6/9
21/03/1759	5/9	8/0	-
19/03/1760	5/10	-	-
28/01/1761	5/9	-	-
08/03/1762	6/7	-	-

The Olive Oil and Raisins Market

Olive oil

Olive oil, which would occasionally replace butter in the rations, seems to have come mainly from Genoa.[50] However, of the thirty tenders presented in the archives, the first thirteen (up to November 1758) did not specify the origins of the olive oil, the stipulation on the contracts being simply for '*olive oil*'. It is therefore highly likely that up to 1758, olive oil may also have originated from Spain, which was geographically closer. However, from 6 November 1758 onwards (apart from on three occasions[51]) and through to the end of the war, the Commissioners' invitations to tender bids were all strictly concerning supplies of Genoa oil only.

The decision to import Genoa oil was most probably due to the tensions (leading eventually to war) between England and Spain, and a study of the price curve in Figure 10.7 would seem to support this theory. In fact, the

50 Baugh, *British Naval Administration*, 417.

51 Bids made on 5 and 9 March and 2 April 1759.

Figure 10.7 Fluctuations in the price of olive oil and of empty jars bought by the Commissioners of the Victualling Board

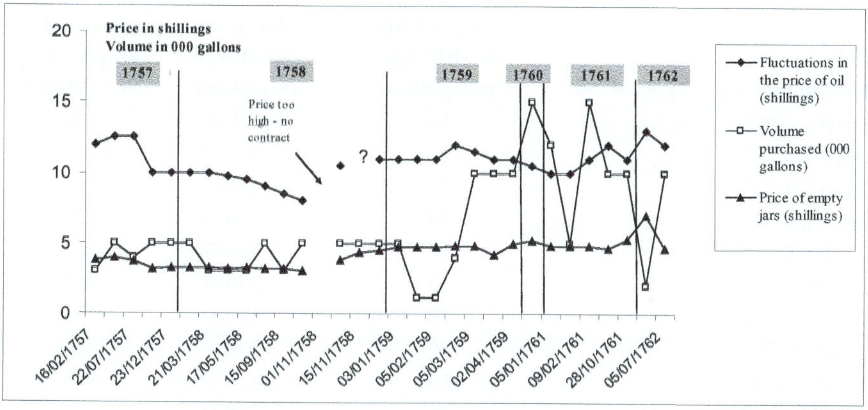

clause stipulating the geographical origin of the olive oil in the contracts coincided with a marked increase in prices. The sudden disappearance from the economic scene of a merchant who had been one of the main suppliers of olive oil – Cornelius Owen – in November 1758, may also have played a role in the increase in prices.

This sector was also controlled by a limited number of traders (see Table 10.13). Even though a total of sixteen merchants took part, at some point during the war, in the invitations to tender bids, a full 84 per cent of the market share went to just two of them: John Le Keux and Cornelius Owen (Le Keux 57 per cent, Owen 27 per cent). Until Owen's retirement in November 1758, that merchant had succeeded in obtaining eight contracts out of the ten tenders he had participated in. John Le Keux had only succeeded in obtaining three contracts up until then, and then began to take advantage of the Board's decision to use Genoa as the sole origin of olive oil importation. As from November 1758, he obtained fourteen of the eighteen available contracts, representing a 78 per cent success rate.

Olive oil prices dropped slightly in the spring of 1757 and then remained steady until November 1758 (see Figure 10.7). The fact that the market had been limited to olive oil from Genoa triggered an increase in tariffs of about 45 per cent, a gallon of olive oil rising from 3s 3d to 4s 9d. These prices remained more or less constant from January 1759 through to April 1761, and then underwent a series of massive increases: from 5s 4d per gallon in October 1761 to 7 shillings per gallon by March 1762. This movement in prices was similar to those that have already been discussed for other foodstuffs, and originated from similar factors: an increase in demand coupled with an increase in lending rates (see Chapter 2).

Table 10.12 Prices and volumes of olive oil and empty jars bought by the Commissioners of the Victualling Board

Bidding date	Type of oil	Price of oil in shillings per gallon	Volume purchased in gallons	Price of empty jars in shillings
16/02/1757	Olive oil	12	3,000	3/10
04/04/1757	"	12/6	5,000	3/11½
22/07/1757	"	12/6	4,000	3/9
19/08/1757	"	10	5,000	3/2½
23/12/1757	"	10	5,000	3/3
06/02/1758	"	10	5,000	3/3
21/03/1758	"	10	3,000	3/3½
17/04/1758	"	9/9	3,000	3/2½
17/05/1758	"	9/6	3,000	3/3
28/06/1758	"	9	5,000	3/2
15/09/1758	"	8/6	3,000	3/2½
27/09/1758	"	8	5,000	3
01/11/1758	"	Price too high – no contract made	-	
06/11/1758	Genoa oil	10/6	5,000	3/10
15/11/1758	"	?	5,000	4/4
01/12/1758	"	11	5,000	4/6
03/01/1759	"	11	5,000	4/9
29/01/1759	"	11	1,200	4/9
05/02/1759	"	11	1,500	4/9
19/02/1759	"	12	4,000	4/10
05/03/1759	Olive oil	11/6	10,000	4/10
19/03/1759	"	11	10,000	4/2
02/04/1759	"	11	10,000	5
25/01/1760	Genoa oil	10/6	15,000	5/3
05/01/1761	Genoa oil or any other oil of equal quality	10	12,000	4/10
12/01/1761	"	10	5,000	4/10
09/02/1761	"	11	15,000	4/10½
08/04/1761	"	12	10,000	4/8
28/10/1761	"	11	10,000	5/4
05/03/1762	"	13	2,000	7
05/07/1762	"	12	10,000	4/8

Table 10.13 Merchants competing for the Royal Navy's victualling in olive oil

Merchants who participated in the bidding

***Merchants acting individually* 16**

(number of tenders)

Barth (2)	William Marshall (4)
Richard Clay (9)	Nicholas Nixon (5)
Thomas Hill (12)	Cornelius Owen (10)
Isaac Hughes (2)	John Potts (6)
William Keeling (2)	Bartholome and William Price (8)
John Knight (3)	Ralph Reed (1)
John Le Keux (26)	Thomas Symmes (1)
Edward London (3)	Daniel Taylor (3)

***Associations* 1**

Messrs John Potts and Daniel Taylor (3)

Breakdown of the contracts awarded

***Number of contracts awarded* 30**

***Suppliers* 6**

Thomas Hill: 2 (shared with Le Keux)	William Marshall: 3
John Le Keux: 17 (2 shared with Tho. Hill)	Cornelius Owen: 8
Edward London: 1	Bartholomew and William Price: 1

Raisins

Smyrna in western Turkey (today called Izmir) appears to have been the Victualling Board's main geographical source of raisins throughout the war – or at least, that is the logical conclusion given the appellation of the raisins – *black Smyrna raisins* – on which the invitations to tender bids were based.[52] However, from December 1761, shortages in this type of grape, which in turn had led to a substantial increase in prices over the past year (see Figure 10.8), encouraged the Commissioners to expand the range of products they bought. They began to put out invitations to tender bids not only for black Smyrna raisins, but also for Alexia raisins and for currants. On 23 April 1762, red Smyrna raisins were also added to the list. However, it seems that this diversification existed more in theory than in practice: in fact, the merchants were rarely able to meet the demand for Alexia raisins and currants, since the prices for these products were even higher than those for black Smyrna raisins (see Figure 10.8).

The raisin market was controlled by a small group of merchants: in fact, only ten traders competed in the twenty-nine invitations to tender bids (see

[52] Unless *black Smyrna grapes* is a generic term for all black grapes. There were no documents to be found which could throw a light on this issue.

Figure 10.8 Fluctuations in the price of raisins bought by the Commissioners of the Victualling Board

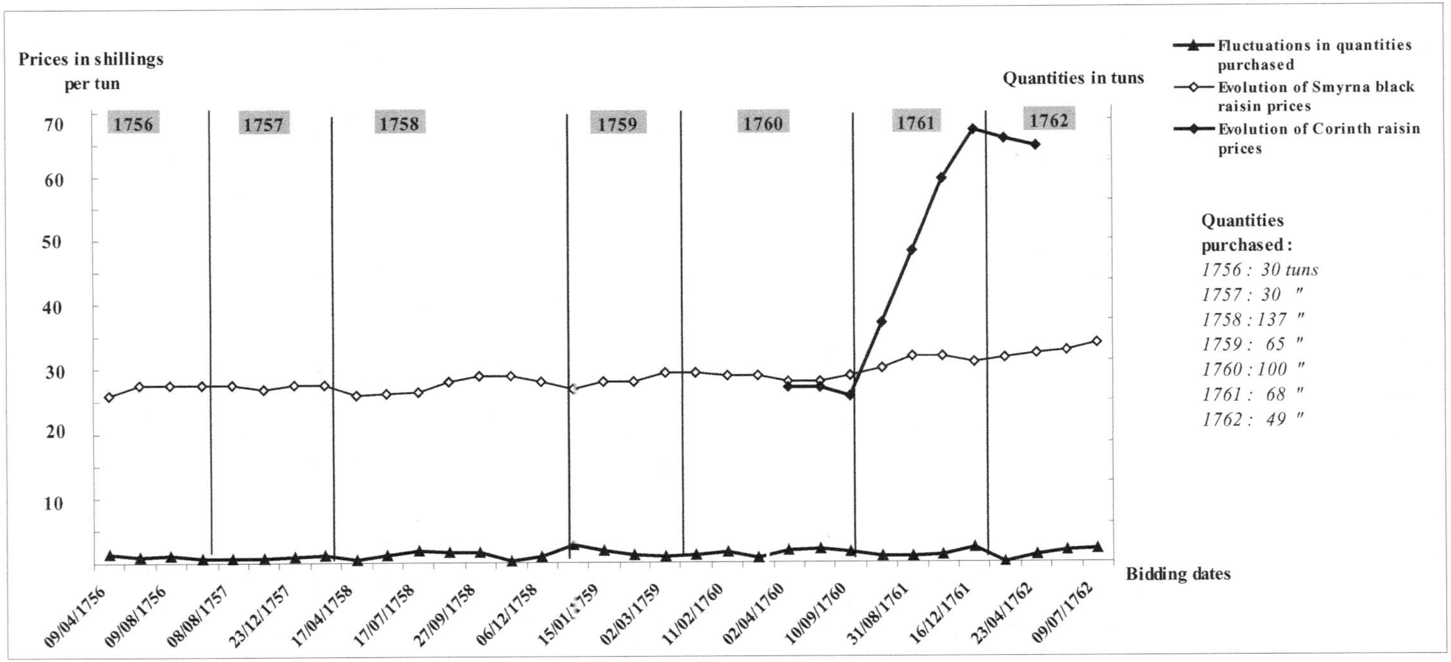

Table 10.15), and the outcome was the complete opposite of what might have been expected from healthy market competition. In actual fact, the ten traders in question had already divided up the market amongst themselves and they all quoted identical rates. As a consequence, the contracts were awarded collectively. This resulted in Francis Harding taking part in all twenty-nine tenders for Navy provisions of raisins, the Stowes took part twenty times, and Henry Bankes sixteen times as an individual and five more in an association (see Table 10.15).

Documents found in the Corporation of London Records Office archives reveal some details about two of these merchants: Abraham Newman (1736–99) and Monkhouse Davison (1713–93). The 1760 *Trade Directory* indicates that Mr. Newman was a corn factor (a corn merchant) with a warehouse in Mark Lane, London. A most interesting booklet was written in 1938 by Owen Rutter entitled *A Brief History of the Firm of Messrs. Davison, Newman and Company* at the request of the company's directors, for the company still existed just before World War II and was a wholesale importer of grapes, figs, pepper, rice, etc. It was then situated in Creechurch Street, EC3, a stone's throw from the Thames. Today, Davison, Newman and Company, although significantly restructured in recent years, continues to exist and operates a wine import business and a tea export business from its London base in Temple Chambers, EC4.

Owen Rutter's booklet reveals that the company originally belonged to a Mr Thomas Rawlinson. When the Lord Mayor of London, Edward Ironside, died in 1753, Thomas Rawlinson was elected as his successor. His duties as mayor obliged him to hand over the management of the company to his partner, Davison, and the firm changed its name to *Rawlinson and Davison Company*. Newman's name was added in 1764. In fact, in 1761, Newman had bought a wholesale company and had married into the Davison family – factors which allowed him to enter the company as a partner. He soon took over the management of the *Rawlinson, Davison and Newman Company*, which played an important role in the Boston Tea Party in 1773.

Upon his death on 8 March 1799, in his Fenchurch Street home, Abraham Newman esquire was

> one of London's richest citizens, having by honest means as a grocer amassed a fortune of £600,000. He had been retired from business for four years but paid a visit to the store every day, eating a lamb cutlet for luncheon there at two o' clock in the company of his successor Mr. Monkhouse. He bequeathed £100,000 to each of his two daughters.[53]

[53] O. Rutter, *A Brief History of the Firm of Messrs. Davison, Newman and Company* (London, 1938), 17.

Table 10.14 The prices of raisins bought by the Commissioners of the Victualling Board

Bidding date	Variety	Price per ton in shillings	Quantity purchased in tuns
09/04/1756	Black Smyrna raisins	26	18
18/06/1756	"	27/6	12
09/08/1756	"	Price too high: no contract made	-
27/10/1756	"	"	-
08/08/1757	"	27/6	10
09/11/1757	"	27	9
23/12/1757	"	27/6	11
23/01/1758	"	27/6	16
23/01/1758	"	26	6
23/01/1758	"	Price too high: no contract made	-
23/01/1758	"	26/6	24
23/01/1758	"	28	20
23/01/1758	"	29	20
23/01/1758	"	29	5
23/01/1758	"	28	12
23/01/1758	"	27	34
15/01/1759	"	28	24
05/02/1759	"	28	15
02/03/1759	"	29/6	12
19/03/1759	"	29/6	34
11/02/1760	"	29	20
03/03/1760	"	29	10
02/04/1760	"	30 (1759 harvest) 28 (1757 harvest)	24
28/05/1760	"	30 (1759 harvest) 28 (1757 harvest)	26
10/09/1760	"	29 (1759 harvest)	20
05/01/1761	"	30	13
31/08/1761	"	32	12
23/10/1761	"	32	15
16/12/1761	Alexia raisins Corinth raisins	31 56	28
27/01/1762	Corinth raisins	62	2
23/04/1762	All raisins	Price too high: no contract made	-
17/05/1762	Corinth raisins	60	21
09/07/1762	Black Smyrna raisins	34	26

Table 10.15 Merchants competing for the Royal Navy's victualling in raisins

Merchants	**Number of participations in tenders**	**Contracts**
Henry Bankes, esquire	18	16
Charles Barrow	2	1
James Bennet	15	13
William Bennet	7	5
Thomas Chitty, esquire	1	1
M. Davidson	18	12
Francis Harding	29	29
Abraham Newman	6	5
George, John and William Stow	22	20
Messrs Henry Bankes and William Bennet	5	5
Merchants 10		
Number of contracts awarded 29		

A third successful bidder was Henry Bankes esquire (bap. 1700–1776), a Dorset landowner, lawyer and MP. These were unlikely qualifications in relation to supplying raisins to the Navy, and it is not improbable that Bankes acted on behalf of the cod fishery merchants and traders of Poole, close to his Corfe Castle constituency and his country seat at Kingston Lacy. Poole was a leading centre of the British cod fisheries, which operated a triangular trade, part of which was the marketing of dried cod in the Mediterranean, with a return cargo of Mediterranean commodities to England.[54]

*

As a general conclusion, Figure 10.9 gives an overview of the evolution in prices of all the products treated in this chapter. Price fluctuations seem to have been under control on the whole, for prices in 1761 were close to those in force in 1757 (which have been taken as the 100 baseline in the graph). The graph illustrates that prices for raisins alone were out of control, having increased as a consequence of the dearth of grapes and also as a consequence of the traders in question collectively exploiting this weakness in the market. In contrast, tariffs for sea beer and land beer and for cheese were lower in 1761 than they had been in 1757. This was all the more remarkable as Suffolk cheese was replaced by Cheshire, Warwickshire and Gloucester cheeses from July 1758 – cheeses of a higher quality, but also a higher cost.

[54] 'Henry Bankes', in L. Namier and J. Brook, eds, *The House of Commons 1754–90* (London, History of Parliament Trust, 1964). I am grateful to Michael Duffy for the suggestion.

Figure 10.9 A comparison of fluctuations in the prices of butter, different types of beer, olive oil, raisins, brine salt and white salt bought by the Commissioners of the Victualling Board (average prices in 1757 are taken as a base of 100)

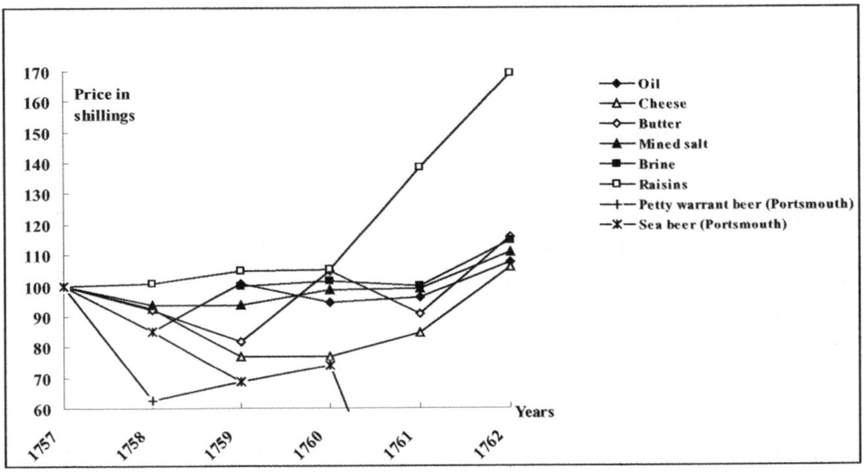

Table 10.16 Comparative fluctuations from 1757 prices of commodities

	1757 base 100	1758	1759	1760	1761	1762
Oil	100	85.12	100.78	94.35	96.19	107.71
Cheese	100	92.31	76.92	76.92	84.62	105.77
Butter	100	92.05	81.82	104.55	90.91	115.91
White salt	100	93.75	94.39	98.62	98.92	110.73
Brine	100	100	100	101.45	100	114.49
Raisin	100	100.64	104.57	105.30	138.50	169.24
Petty warrant beer	100	-	62.33	68.75	-	-
Sea beer	100	-	68.82	74.14	-	-

To the above examples can be added the replacement of malt spirits by French or Flemish brandy and of Malaga wine by Oporto wine – all testimony to the desire of the Commissioners of the Victualling Board to constantly improve the quality of rations throughout the Seven Years War. The fact that these rations also increased in quantity over this period underlines the success with which the Commissioners were able to control costs. All of these points reveal the increasing efficiency with which the British Navy was victualled in this vital period in its history.

Conclusion

> The sea is English in inclination; she does not like France; she destroys our ships; she silts up our ports.
>
> Jules Michelet, *Histoire de France*, Preface, 1869

The picture painted of pre-1790s victualling by maritime historiography is as gloomy as it is erroneous. Almost every publication has quoted the same extracts from William Thompson's *An Appeal to the Public in Vindication of Truth and Matters of Facts*, in which the author, who was employed as a master cooper by the Victualling Board in the 1740s, describes the deplorable conditions and methods in which and by which food provisions were conserved. On the basis of Thompson's statements, it is only too easy to hastily discredit the work of the Commissioners and to judge them as having been, at best, incompetent and negligent, at worst, corrupt. This judgement provided a facile and logical explanation for the high mortality and morbidity rates in the Royal Navy right up to the end of the eighteenth century. Yet the fact that William Thompson's two earlier works were equally exaggerated,[1] not to mention his dismissal by the Victualling Board in January 1745 for having overseen the preservation of provisions which turned bad, should perhaps have warned future historians to be wary of his accusations.

In his admirable work, *British Naval Administration in the Age of Walpole*,[2] Professor Daniel A. Baugh rendered justice to the Victualling Board's work up to the War of the Austrian Succession; similarly, the research in which the author of this present book has been engaged also demolishes a number of popular myths. In fact, the Royal Navy's victualling during the Seven Years War could not have been more efficient, and played an enormous role in the '*ringing of the bells of victory*' (to paraphrase the title of R. Middleton's book[3]).

The Board's success can easily be measured by an examination of financial data. The annual cost for the victualling of 70,000 to 85,000 men with rations of superior quality during the Seven Years War was almost identical to

1 W. Thompson, *The Royal Navy Men's Advocate* (London, 1757); *The Case of William Thompson* (London, 1758).

2 Baugh, *British Naval Administration.*

3 R. Middleton, *The Bells of Victory: The Pitt-Newcastle Ministry and the Conduct of the Seven Years' War, 1757–1762* (Cambridge, Cambridge University Press, 1985).

Figure A A comparison of fluctuations in the prices of oats, butter, wheat, flour, cheese, groats, malt and peas bought by the Commissioners of the Victualling Board (average prices in 1756 are taken as a base of 100)

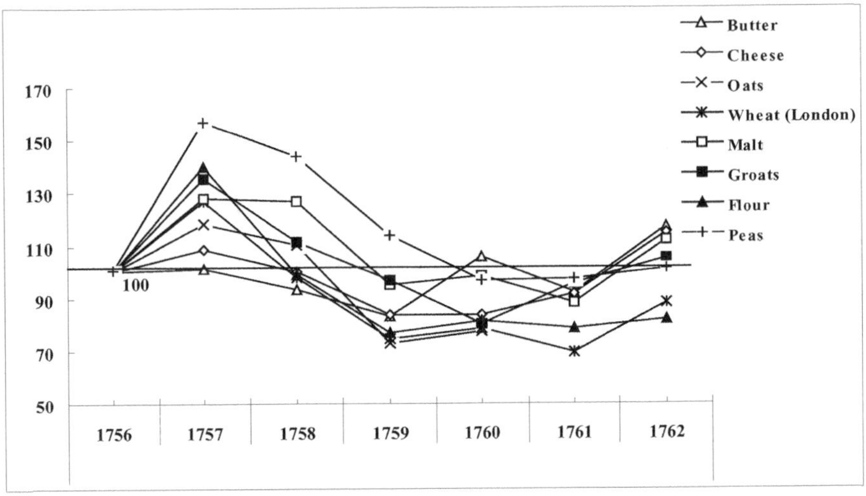

Figure B A comparison of fluctuations in the prices of different types of meat bought by the Commissioners of the Victualling Board (average prices in 1756 are taken as a base of 100)

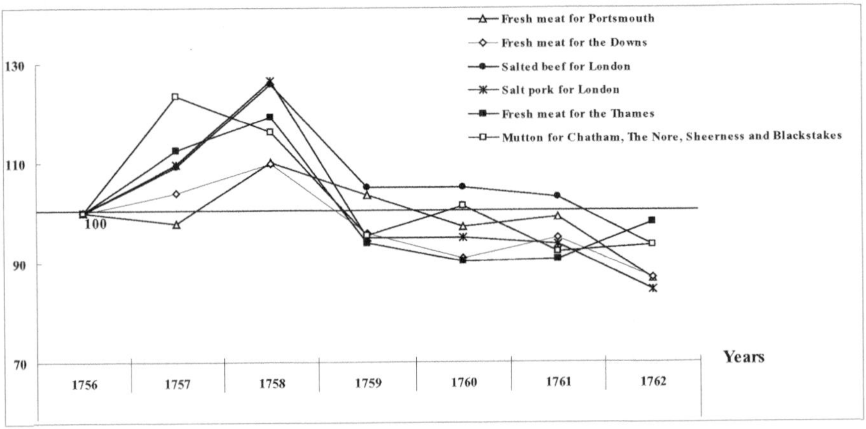

the cost of victualling 40,000 to 50,000 men during the War of the Spanish Succession. Victualling expenditure for 70,000 men serving in the Navy in 1758 was £913,905, a lower figure compared to that of £942,879 in 1710. By 1759, the number of men serving had risen to 84,464, and yet expenditure exceeded by only 10 per cent the sum that the Treasury had spent in 1710: namely, £942,879 in 1710 compared to £1,053,163 in 1759.

This very positive development was the result of the Board's improved control over the prices of the different foodstuffs that made up the men's daily rations. Generally speaking, after the sharp rise in prices that occurred in 1757 – increases of between 1 per cent and 57 per cent with respect to average prices in 1756 (see Figures A and B) – prices then began to fall again in 1758, even though the total number of men who needed victualling had grown. Prices for various goods in 1759, 1760 and 1761 were below the average asking price for the same articles in 1756. It was not until 1762 – a year which was affected by the financial crisis of the autumn of 1761, the bad harvest and cattle disease,[4] the psychological trauma following the declaration of war between England and Spain in January 1762, and the Navy's growing delays in paying its creditors – that average prices overtook those of 1756, without, however, attaining those of 1757, except for the following products: brine salt, white salt, olive oil, butter, cheese and raisins. It should also be borne in mind that the cost of butter and cheese never exceeded the 1757 prices, due to the control exercised over these prices from the very beginning of the war. Price increases regarding brine salt and white salt were the result of the restrictions placed on the import of Irish meat from December 1761, which led to a huge increase in the demand for English salt. The record prices attained for olive oil and especially for raisins in 1762 were the result of their profile: imported products that impeded the competitive market system. In fact, of all the products examined in these chapters, the only product that the Victualling Board was truly unable to control was raisins. Grape prices increased exponentially from 1761 as a direct result of shortage and the traders' willingness to come to mutual agreements and understandings amongst themselves.[5] The meat sector experienced some limited fluctuations in price, but, without exception, prices in 1762 were lower than those quoted in 1756.

There was also notable progress in terms of distribution. Food shortages, which had been so common during the wars of the League of Augsburg and Spanish Succession and which had been responsible for putting a strain on the fleets' offensive powers, became rare in the 1739–48 war, and altogether

4 Macdonald, 'A new myth of naval history?', 175, 181.

5 And possibly because of the threat to trade with the Mediterranean due to the war with Spain.

exceptional during the Seven Years War. The disappearance of the *short allowance* system is, in itself, significant.

The low quantities of badly preserved food (less than 1 per cent loss, with the exception of stockfish) indicate, contrary to popular opinion, the degree to which the Victualling Board had succeeded in perfecting its food conservation techniques. Meat losses were very low (0.06 per cent for beef, 0.03 per cent for pork) and illustrate how far techniques had progressed since the War of the Spanish Succession, in which cured meat had been particularly substandard: 'The salting of beef and pork was done very crudely; a high proportion had to be condemned after being on board but a few months.'[6] This vast improvement in the quality of produce, as reflected in the food preservation statistics, undoubtedly played a very positive role in the morale, discipline and health of the men.

It is also a disappointment that focus on the Board's sparing use of lemons before the end of the eighteenth century should too often have eclipsed the immense efforts undertaken in protecting their men against scurvy. In fact, the Seven Years War can be considered as a period of total change in terms of diet management due to the fact that there were so many innovations in the fight against scurvy.

The medical profession at first believed that cured foods were at the root of the problem because they were difficult to digest; as a result all means possible were undertaken to try to provide fresh meat rather than salted beef or pork. In terms of scurvy, the effort was ineffective because fresh meat is not a source of vitamin C. However, the initiative deserves to be underlined and applauded, especially since it reveals the Board's determination to eradicate the disease and contradicts maritime historiography's view of the Board as negative and complacent.

The increase in consumption of fresh vegetables during the Seven Years War was in itself catalysed by another innovation – the introduction of 'portable soup'. Both these factors had very positive effects in helping in the eradication of scurvy. The establishment of a group of supply ships in the second quarter of 1759, which had the role of victualling the main battle fleet (the western squadron) with live animals and fresh vegetables, opened a 'new chapter in the history of the victualling of the Royal Navy'[7]. Through its vital impact upon strategy, the improvement in the logistical system was without dispute the determining factor in enabling British success in European waters.

In a wider perspective, N.A.M. Rodger has already shown that mortality

6 Merriman, *Queen Anne's Navy*, 250.

7 Gradish, *The Manning of the British Navy*, 168.

and morbidity rates fell in the 1757 to 1762 period.[8] In this way, the parallel and complementary actions, from a medical and administrative point of view, of the Victualling Board and the Sick and Wounded Board permitted the Royal Navy to almost control the one factor which in the past had dictated the success or failure of operations, namely, the health variable.

In retrospect, can it be claimed that England's early agricultural revolution (analysed by P. O'Brien and E.L. Jones[9]) provided an especially stable setting? By the seventeenth century, England had already overcome famine and reduced the effects of bad harvests so that they no longer had disastrous demographic consequences.[10] More closely concentrated properties and larger stocks of productive assets (especially animals) encouraged higher yields than on the continent.[11] Added to this were other favourable factors such as the country's insularity, and its relatively small size which meant that no region was too far removed from any other. All in all, these factors were positive points in the country's advanced agrarian market economy. Also, in contrast to France, a subsistence economy had virtually disappeared by the seventeenth century.

The naval administration, in particular, and the state more generally, had to exploit this favourable background. The efficiency of the victualling system that was progressively established under the jurisdiction of the Victualling Board is proof of this. Once again, British methods differed considerably from French ones. The victualling of French warships had been the responsibility, since 1669, of one sole entrepreneur, the Victualler General of the French Navy. The Victualling Board, on the other hand, was acting on experience from a decision first taken in 1654 when merchants were invited to tender bids in response to the Victualling Board's offers.

The Commissioners responsible for the principal naval ports whose provisioning was directly under the jurisdiction of the Victualling Board (London, Portsmouth, Chatham, Plymouth and Dover) chose to pay close attention to the realities of the market economy by employing the invitation to tender bids system, and breaking down the goods required into seventy-one distinct

8 Rodger, *The Wooden World*.

9 P.K. O'Brien, 'Agriculture and the home market for English industry, 1660–1820', *English Historical Review*, 100 (1985), 773–99; and E.L. Jones, *Agriculture and Economic Growth in England, 1650–1815* (London, Methuen, 1967).

10 J.-P. Poussou, in *Histoire Economique et Sociale du Monde*, ed. P. Léon, vol. II, *Les hesitations de la croissance 1580–1740*, 49.

11 Yields of oats were particularly impressive, increasing by 40 per cent during the eighteenth century: M. Turner, 'Agricultural productivity in England in the eighteenth century: evidence from crop yields', *The Economic History Review*, XXXV (Nov. 1982), 489–510.

categories and services. Apart from the very beneficial effects this competitive system had for price control, it had the added advantage of producing highly specific contracts which in turn required highly specialised suppliers. These suppliers were in a far better position to carry out their tasks than the French-style Victualler General who was responsible for all kinds of victualling. As a consequence, food administration was much more efficient. In fact it would have been impossible to terminate a Victualler General contract for one aspect of any specific problem, without harming the victualling of the French Navy as a whole. However, the British system could use such sanctions due to the specialised nature of its contracts, and it was able to reject tenders in any commodity if the price was felt to be too high. The threat of sanctions that constantly menaced the suppliers, plus the limited duration of their contracts, put the Victualling Board in a position of superiority whereby they could impose a series of lesser obligations – such as asking for samples or imposing guarantees – plus-points which did not cost the Crown a penny.

The Victualling Board's control was rendered even more efficient by the growth and development of London. The capital played a central role as the major British consumer market and financial centre during the agricultural and industrial revolutions (cf. F.J. Fisher, F. Crouzet and J.-P. Poussou[12]), and thus the most important merchants continued to live and operate in its centre, or at least in the suburbs. This was invaluable to the Commissioners since it enabled them to offer and draw up contracts directly without having to use the services of an agent. In sum, the Commissioners were able to benefit from a multi-structured market and to take full advantage of the competitive system.

Apart from the product-by-product specialisation that took place in the victualling of the main bases, there was also geographical specialisation in terms of the second-tier metropolitan ports and the outlying ports overseas. A single merchant in each of these ports had the task of meeting the Navy's needs. This manner of proceeding in ports that did not achieve the same level of activity as London, Plymouth, Portsmouth, Chatham or Dover relieved the Victualling Board of the task of managing warehouses which would otherwise have been put under the control of an Agent, who in turn would have been assisted by a number of employees – a cost which was not economic in light of the limited number of rations involved. The system had the added

12 Fisher, *London and the English Economy 1500–1750*; F.J. Fisher, 'London as an engine of economic growth', in P. Clark, ed., *The Early Modern Town* (London, Longman, 1976), 205–6; F. Crouzet, *De la supériorité de l'Angleterre sur la France, l'Economique et l'Imaginaire XVIIème–XXème siècles* (Paris, Perrin, 1985); J.-P. Poussou, 'Les Villes anglaises du milieu de XVIIe siècle à la fin du XVIIIe siècle', in Poussou, ed., *Les Villes en Europe Occidentale* (Paris, CDU, 1983), vol. II, 1–212.

advantage of allowing the Victualling Board to diversify the sources of its provisions, which led directly to better management of requirements and better financial management, since it was also possible to make the most of growing regional economies. For example, when the Commissioners decided to give up direct management of Navy victualling in Jamaica and, instead, to entrust the victualling to a contractor, they made savings of approximately 14 per cent.

The Navy both took advantage of the development of commercial capitalism and contributed to it, as can be seen from the example of John Biggin's remarkable breakthrough during the war. Biggin's company was established in England, Ireland and North America, and these three centres of activity – helped by the routes the ships took in the Atlantic, plus the fact that his two sources of provisions were Ireland and South Carolina – allowed Biggin to dominate the victualling market in the intermediary bases such as Lisbon, Oporto, Madeira and Gorée (today's Senegal), and to establish himself in the West Indian area. Further advantaged by the intercontinental nature of these markets and by the development of naval operations carried out by Britain in the West Indies and Canada, John Biggin went from strength to strength during the last years of the war.[13]

The spreading of similar networks (such as those created by John Allen, Anthony Bacon, Auguste Boyd, Messrs Evory and Clements, Samuel Fouchet, Sir Alexander, Robert and William Grant, Messrs Nesbitt and Franks and James Riddell) reinforced the competitive aspect which, although burdening regional merchants, was of the greatest benefit to the Victualling Board which was now able to take advantage of increasingly lower prices. Indeed, in the eighteenth century, Great Britain and its colonies comprised the largest free-trade zone in the world.[14]

The Commissioners of the Victualling Board aimed at maintaining a competitive market at all times. Their response when meat prices shot up in 1758 was to diversify their sources of meat provision; consequently they offered invitations to tender bids for Irish meat throughout the remaining

13 Biggin undoubtedly did well out of the war. In 1764 he purchased the manor of Cosgrove Priory, Northants, and after his death in 1775, his widow Eleanor received in payment of a still outstanding debt Anthony Bacon's manor and newly built house at Higham Hill, Essex, which the latter had also bought with his war profits in 1764. www.histfam.familysearch.org (John Biggin); 'The Manor of Cosgrove after 1487', in P. Ridden and C. Insley, eds, *A History of the County of Northampton, vol. 5: The Hundred of Cleley* (London, Victoria County History, 2002), 77–98, D. Lysons, *The Environs of London, vol. 4: Counties of Herts, Essex and Kent* (London, 1796), 209. I am grateful to Michael Duffy for this information.

14 K. Berrill, 'International trade and the rate of economic growth', *Economic History Review*, new series, XII, no. 3 (April 1960), 358.

years of the war – a move that allowed them to keep their meat prices under control. This is why the 747 contracts for the different provisions that have been examined in this book reveal, without the slightest doubt, that there was no question of complicity or of tacit agreements between the Commissioners and any of the merchants. Contracts were issued openly, without the slightest veil of secrecy, and it was quite simply the most suitable trader who won the contract.

The huge development of infrastructure in London, Portsmouth and Plymouth, and to a lesser extent Chatham and Dover, transformed these yards into major victualling manufacturers: 1,360 personnel were employed in these yards in 1761. This evolution was not the result of an ideological premise aimed at increasing the state's power, but resulted from the simple observation that it was much easier to judge the quality of provisions in their raw state rather than once they had been fully or partially processed; in fact, the Victualling Board's intention was not to replace the market, but to control it. Further, the fact that foodstuffs were processed under their supervision allowed the Commissioners not only to cut costs, but also to ensure that strict, new procedures were followed, which helped to ensure better preservation of victuals.

This very positive development was not simply due to the evolution of British commercial capitalism or even to the fact that the Victualling Board was able to stimulate the market economy. It was also due, in part, to the Admiralty's backing. The extraordinary prestige of Lord Anson, First Lord of the Admiralty, coupled with his highly respected position and connections within the government, conferred tremendous power and authority on the Admiralty Board.[15] This reinforced the Board's control over the various naval administrative sections and ensured their mutual cooperation and coordination – aspects that had been missing in the past, to the detriment of the good running of the Navy. The author has elsewhere illustrated the extent to which the arming of ships under the auspices of the Ordnance Board was often delayed due to breakdowns in communication between that department and the Navy Board, thus frequently delaying the departures of convoys during the 1739–48 war.[16] Also not to be forgotten, but often overlooked, is the inspirational personal influence of Anson, the hero (and a scurvy victim) of the round-the-world voyage of the 1740s. It was thanks to him that attention was paid to issues of health and hygiene, obliging the Commissioners of the Sick and Wounded Board and the Victualling Board to be more enterprising.

15 Rodger, *The Admiralty*, 59–68. Anson's father-in-law was Lord Chancellor Hardwick, closest ally and confidant of the First Lord of the Treasury, the Duke of Newcastle.

16 Buchet, 'L'Ordnance Board et la logistique en armes pour les flottes anglaises aux Antilles (1689–1763)'.

Without Anson's dedication and support, it is unlikely that the funding for expensive innovations such as 'portable soup', the replacing of Suffolk cheese by other more expensive varieties and the sending out of merchant ships laden with live animals and fresh vegetables in order to victual the western squadron, would have occurred as quickly and efficiently.

This level of efficiency could not have been attained without Britain's unprecedented financial support for its Navy. It is largely thanks to the fact that the Navy did have the financial means to create victualling yards and to pay back its creditors within reasonable time-limits, that the Victualling Board was able to establish a reliable and coherent logistical system, based on clearly defined procedures in terms of the buying and stocking of foodstuffs. In contrast, during the War of the League of Augsburg, the Commissioners had been financially unable to install a competitive structure in which to operate the buying process, and indeed were forced to beg for advances from the suppliers, thus putting them at the mercy of the very men they were trying to control. The state's plan for victualling included optimising the contractor system under the jurisdiction of the Victualling Board, which involved financial means that seventeenth-century governments simply did not possess.

These capital investments were largely recovered, in that the cost of feeding one man during the Seven Years War (including infrastructure costs) diminished somewhat in relation to the cost during the 1739–48 war, and diminished greatly in relation to the War of the Spanish Succession.

Furthermore, these financial investments also permitted the Navy to establish important naval arsenals along a new east-west defensive line on the south coast, which progressively replaced the old north-south line on the east coast dating back to the Anglo-Dutch wars. Portsmouth, Plymouth and to a lesser extent Dover underwent major development, and Portsmouth and Plymouth followed London (Deptford and Woolwich), Chatham and Sheerness in having a number of dry docks at their disposal. Not to be left out either were the far-distant bases, with, for example, the construction of a new naval yard at Port Antonio in Jamaica – the biggest development of its type in the 1714–39 period; or the expansion of the bases in Gibraltar, Port Mahon, Halifax and Antigua – where the yard at English Harbour was built and was employing 347 employees by January 1762.[17] Consequently, the Royal Navy had the means by which to keep all its units operational, and at a far superior level than those of the French or Spanish navies.

Similarly to the Victualling Board or the Navy Board, the Sick and Wounded Board also benefited from generous budgeting. Hospitals were built

17 C. Buchet, 'Révélateurs d'une determination coloniale: les arsenaux anglais de l'espace caraïbe au XVIIIe siècle', *Revue Histoire, Economie et Société*, 1 (1997), 125–49.

in Portsmouth and Plymouth and the number of hospital ships increased. These measures, alongside those adopted to achieve better hygiene and sanitation and to train medical staff, contributed to an improvement in care and disease prevention, whilst at the same time reducing the number of deserters.

The Navy's financial backing was due to Members of Parliament who believed in the common good to be gained from having a powerful Navy. After the 1688 Revolution which proved to be so favourable to the House of Commons, and the disaster at Beachy Head in 1690, the threat of a French landing on English soil in order to restore James II to the throne seemed all too possible. Parliament's response in the autumn of 1690, therefore, was to order the construction of twenty-seven vessels. Naval development became a priority, almost an obsession, and helped to reinforce the sense of pervading mercantilism that was infiltrating the House of Commons. The powerful West Indian planters lobby (known as the West Indian Interest), which was very strongly represented in Parliament throughout the eighteenth century, saw the Navy as the means by which to conquer and keep lucrative colonies.

Due to a financial structure that was much more sophisticated in Britain than in France, this political determination was given substance. The Bank of England in London had been established since 1694, the Stock Exchange and country banks were in full expansion, and paper money was used widely without creating inflation. L. Neal and C.H. Wilson have highlighted London's increasing attractiveness as a financial market for holders of capital in the United Provinces of the Netherlands.[18]

It is undoubtedly true that Britain benefited from an efficient fiscal system linked to a strategy of long-term borrowing, the whole rendered stable by careful and sensible management of the public deficit. Its counterpoint was illustrated by the sluggishness of the Versailles administration, entangled in financial problems created by a poor state operating in a rich country in which the elite refused to finance the necessary costs (cf. Jean Meyer's *Le Poids de l'Etat*). Consequently, as revealed by the European survey of 1759 and contrary to popular thought, France of the *ancien régime* was not overtaxed, whilst Britain taxed both its elite and the people at a much higher rate: nine to ten pounds were deducted annually per person in France under Fleury, whilst in Britain at the same time a sum of twenty pounds annually per person was deducted. Consequently, even though British war expenditure equalled that of France – shown by Paul Butel and Michel Morineau to have

18 L.D. Neal, 'Interpreting power and profit in economic history: a case study of the Seven Years' War', *Journal of Economic History*, XXXVII, no. I (March 1977), 20–35; C.H. Wilson, *Anglo-Dutch Commerce and Finance in the Eighteenth Century* (Cambridge, Cambridge University Press, 1966); C.H. Wilson and A. Carter, 'Dutch investment in eighteenth century England', *The Economic History Review*, XII, no. 3 (April 1960), 434–9.

been 1.8 billion pounds between 1756 and 1763 – England was nonetheless able to cover 81 per cent of this sum by having recourse to loans and without having to confront an increase in interest rates. This expedient had already been adopted in the past – during the War of the Austrian Succession, for example, in which the government had financed the cost of its armed forces by taking out loans with interest rates not exceeding 3 to 4 per cent (these rates represented half of those charged during Marlborough's time). The American Revolution had disastrous financial consequences for the French monarchy, and yet Paul Kennedy has calculated that the national debt was appreciatively similar in England and in France. The difference lay in the fact that annual interest repayments were twice as high in France.[19]

Perhaps the secret of the British Navy's success, of the link between the country and its Navy, can be found in the large credits made available to the Navy (60 per cent of the armed forces budget) throughout the eighteenth century. Perhaps these large subsidies rendered it more visible to the public, who took a greater interest in it and demanded to see an efficient, successful operation. In any case, the Navy did become an object of scrutiny. It represented not only national security, but also wealth and power. Serving in the Navy was considered to instil a sense of self-sacrifice into a young man and to be an initiation into '*real*' life. The men who served in it were heroes, with a certain social standing that was missing for their French counterparts. Of the thirty admirals serving during the Seven Years War, twenty of them were also Members of Parliament.

The building of a powerful Navy was surely one of the major forces behind British economic development, as seems to have been the case for the United Provinces of the Netherlands in the seventeenth century and the United States in the nineteenth century.[20] In conclusion, is it not probable that this interaction between the Navy, economy and society – an interaction that was often absent in the French context – was indeed the explanation for France's economic tardiness, and not, as Max Weber suggests, its catholic culture?

19 P. Kennedy, *The Rise and Fall of the Great Powers* (London, Unwin, 1988), 84.

20 The development of the Navy has had important consequences in technical history. The overuse of wood, exacerbated by naval construction, accelerated the move to coal, which contributed to the development of the steel industry. Sails, ropes, boilers, all required in great quantities for a fleet, stimulated economic activity. There is no transport industry that did not benefit from the expanding naval industry. Cf. R. Jenkins, 'Copper smelting in England', *Transactions of the Newcomen Society*, XXIV (1943–5), 73–80; A.H. John, 'War and the English economy, 1700–1763', *Economic History Review*, new series, 7 (1954–5), 329–44.

APPENDIX 1

Ordinary Charge of the Victualling Board in 1747

London	
Seven Commissioners at £400	£2,800
house money to four at £65	260
Secretary	200
4 clerks	240
Cashier	150
4 clerks	200
Accountant for Cash	120
7 clerks	330
Clerk for Assignment of Bills and to Cheque on the Treasurer's Accounts	80
2 clerks	90
Accountant for Stores	120
5 clerks	250
Clerk of the Issues	80
1 clerk	50
Clerk for keeping a Charge on Pursers' Accounts	60
1 clerk	50
Clerk for bringing up Stores in Arrears	60
1 clerk	50
Chief Clerk for stating and passing Pursers' Accounts	100
1 clerk	50
Master Cooper	80
2 clerks	90
Clerk at Tower Wharf for Casks, Stores, and Hoops	50
Wharfinger at the Redhouse, Deptford	50
Clerk of the Cuttinghouse at Tower Hill	80
house money	30
2 clerks	80
Master Butcher	40
Surveyor of Oxen and Hogs	40
Clerk of the Bakehouse at Tower Hill	100
house money	30
2 clerks	90
Clerk of the Mills and the Bakehouse at Rotherhithe	80
1 clerk	40
Clerk of the Brewhouse at Hartshorn	60
3 clerks	110
Clerk of the Cheque at the Brewhouse	80
3 clerks	120

Master Brewer	57
messenger	20
Chief Clerk of the Short Allowance Office	50
2 clerks	80
Surveyor of the Works	60
Hoytaker (£60 plus £20 house money)	80
Master Distiller	80
Total London salaries	**£7,017**
Rent, hoys, repairs, stationery, taxes, contingencies, etc.	9,167
Total London Office	**£16,184**
Portsmouth	
Agent	200
Storekeeper	80
Clerk of the Cheque	80
Master Cooper	60
Minor Clerks, rent, hoys, repairs, stationery, etc.	2,415
Total	**£2,835**
Plymouth	
Agent	200
Storekeeper (£60 plus £10 house money)	70
Clerk of the Cheque	60
Master Cooper (£60 plus £10 house money)	70
Master Brewer	50
Other charges (similar to Portsmouth)	2,696
Total	**£3,146**
Chatham	
Agent	200
Storekeeper	50
Master Cooper	47
Other charges	963
Total	**£1,260**
Dover	
Agent	200
Storekeeper	50
Chief Clerk	50
Other charges	1,965
Total	**£2,265**
Gibraltar	
Agent (£200 plus £50 table money)	250
Chief Clerk (£60 plus £30 table money)	90
Clerk (£50 plus £25 table money)	75
Cooper	39
Other charges	127
Total	£581

Port Mahon		
Agent (£200 plus £50 table money)		250
5 Clerks (£50 plus £25 table money each)		375
10 Coopers (at £39)		390
Master Furner		50
3 Bakers (at £35)		105
Other charges		650
	Total	**£1,822**

Jamaica		
Agent		300
Chief Clerk		120
Clerk		100
2 Clerks (at £80)		160
Master Cooper		80
Cooper		62
Other charges		1,126
	Total	**£1,948**

Kinsale		
Agent		60
Storekeeper		60
Other charges		232
	Total	**£352**

	Total Ordinary Charge of the Victualling Office	**£30,393**

Source: Baugh, *Naval Administration in the Age of Walpole*, 55–7, based on ADM 110/15, 28 January 1747.

APPENDIX 2

Commissioners of the Victualling Board 1755–63

Inspector of the business of the Accountant General, the office for keeping a check on the Treasurer, the Short Allowance Office, and the Muster Master of the Workmen and Labourers who make up the Pay Books.

James Wallace	February 1746

The Manager of the Secretary's Office, the Office for keeping a Charge on Pursers, the Office of the Clerk of the Issues, and the Office for Comptrolling Accounts and Stores.

William Jenkins	July 1747 – March 1759
John Rule	March 1759 –

Inspector of the Bakehouses and Mills

Sir Francis Styles	June 1738 – February 1762
Jonas Hanway	March 1759 –

Inspector of the Brewhouse

Francis Vernon	June 1747 – March 1759
Tyringham Stephens	March 1761 –

Inspector of the Cooperage

Sir Roger Burgoyne	July 1752 –

Inspector of the Business of the Hoytaker

Captain Robert Pett	February 1755 –

Inspector of the Business of the Cutting House

Horatio Townshend	June 1747

Source: Gradish, *The Manning of the British Navy*, 27.

APPENDIX 3

The Structure of British Naval Administration

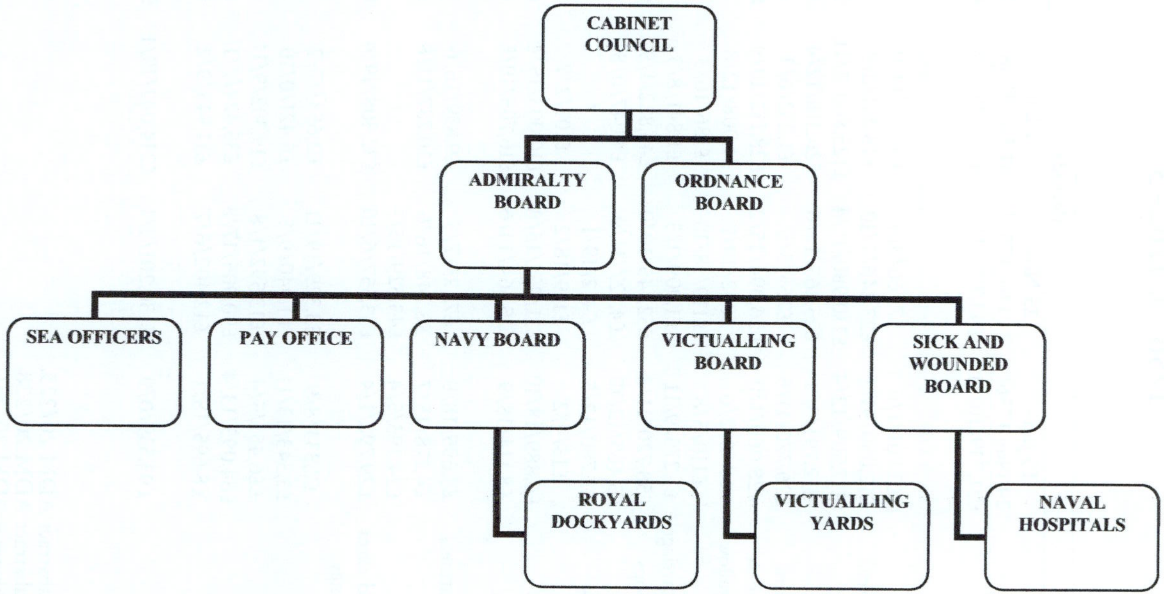

APPENDIX 4

Itemised Distribution of Victualling Board Expenses, 1756–9, 1762–3

Items	Periods			
	13 Jan– 24 Nov 1756[1]	**25 Nov– 31 Dec 1757**[2]	**1 Jan– 31 Dec 1758**[3]	**1 Jan– 31 Dec 1759**[4]
Balance of Victualling Accounts	£21,739/10/11	£43,775/2/3	£56,306/9/3	£50,575/15/3
Bavins	£3,801/1/6	£7,166/3/4	£5,902/14/0	£5,258/7/1
Biskett and bread	£12,050/91/1	£23,246/7/10	£18,241/2/3	£29,555/2/4
Beef and suet	£117,509/12/4	£110,980/11/4	£125,800/8/11	£164,181/5/8
Beer	£12,791/3/3	£25,265/12/6	£26,318/12/4	£9,194/12/11
Brandy and rum	£26,770/11/5	£765/3/5	£1,252/6/3	£18,334/5/6
Butter and cheese	£28,060/1/0	£26,094/15/2	£28,112/10/9	£38,526/12/9
Candles and tallow	£940/2/6	£842/19/7	£1,008/12/8	£887/7/8
Canvas	£110/5/8	£110/18/10	£186/16/3	£192/12/3
Cartage and carriage	£1,275/9/11	£1,909/1/3	£1,841/8/3	£2,349/13/1
Casks and hoops	£58,720/11/3	£52,145/10/0	£43,822/1/9	£35,847/4/10
Coals	£3,465/12/10	£4,127/11/9	£4,047/0/8	£4,872/16/11
Contingencies	£2,780/12/5	£412/19/1	/	/
Cordage	£13/4/2	£119/18/2	£103/2/7	£205/1/4
Flour	£6,889/18/10	£11/251/15/8	£10,003/15/4	£12,973/16/9
Freight and demurrage	£8,114/18/9	£15,364/14/6	£16,705/19/5	£19,322/15/1
Grotts and oatmeal	£3,695/18/9	£7,123/7/5	£9,839/16/6	£16,926/3/0
Hay	£4,178/15/7	£8,248/16/6	£10,787/11/8	£20,710/19/2
Malt	£24,393/5/4	£43,629/15/7	/	£46,149/5/6
Necessary and extra necessary money	£29,792/1/4	£33,557/6/10	£32,308/19/4	£37,861/17/10
Oats	£2,333/8/9	£3,329/19/11	£2,653/17/2	£1,133/1/7
Oil	£5,343/13/11	£1,940/9/5	£5,421/0/10	£15,522/6/6
Peases	££6,451/5/4	£13,592/1/8	£19,302/9/11	£17,275/1/8
Pork	£64,095/13/4	£50,905/17/8	£35,424/7/1	£101,389/3/0
Promiscuous payments[5]	£8,695/18/1	£12,402/6/7	£11,545/9/2	£15,028/14/9
Provisions[6]	£93,539/0/9	£130,201/3/1	£214,042/19/1	£204,940/4/11

1 Register abstract ADM 20/225.
2 Register abstract ADM 20/228.
3 Register abstract ADM 20/231.
4 Register abstract ADM 20/235.
5 Payments without a specific order or precedent.
6 Payments to contractors at the second-rank home ports and the overseas ports for goods provided.

Rent	£13,305/3/3	£754/10/7	£1,059/6/8	£1,057/10/8
Salary	£6,570/19/7	£4,704/7/1	£6,587/13/9	£7,623/9/3
Salt	£15,217/16/4	£12,238/1/10	£11,912/10/8	£9,372/3/5
Stationery ware	£65/18/8	£1,229/14/4	£1,019/10/8	£834/9/5
Staves	£15,045/10/5	£15,614/0/1	£12,727/11/2	£5,111/16/4
Vinegar	£1,537/13/4	£2,304/9/1	£2,740/19/9	£3,348/10/3
Wheat	£3,898/15/10	£113/405/8/7	£108,356/17/10	£69,164/19/4
Wine	£5,334/0/0	£16,475/17/3	£17,701/3/5	£13,018/8/5
Workmanship	£9,380/17/8	£20,637/13/11	£17,701/3/5	£13,018/8/5
Extra payments	£1,416/12/6	£350/0/0	£300/0/0	£300/0/0
Wages	£37,518/6/9	/	/	£62,638/15/1
Total	**£697,406/11/1**	**£816,220/2/1**	**£913,905/19/4**	**£1,053,163/19/1**

Items[7]	**Periods**	
	2 June–31 Dec 1762[8]	**1 Jan–31 Dec 1763**[9]
Balance of Victualling Accounts	£42,862/15/9	£7,666/6/10
Bavins	£4,341/0/4	/
Biskett and flour	£15,427/16/2	£2,002/0/4
Beef and suet	£65,928/5/7	£539/6/7
Beer, hop and malt	£11,238/5/7	£322/4/5
Brandy and rum	£5,259/16/6	£183/12/0
Butter and cheese	£21,249/2/1	£629/2/4
Candles and tallow	£911/14/4	/
Canvas	/	/
Cartage and carriage	/	£42/0/10
Casks, hoops and staves	£22,695/4/2	£7/0/0
Coals	£2,828/4/4	£226/8/2
Contingencies	/	/
Cordage	/	/
Flour	(*cf. Biskett and flour*)	(*cf. Biskett and flour*)
Freight and demurrage	£26,928/4/4	£226/8/2
Grotts, oatmeal, oates	£4,716/5/4	£725/0/8
Hay	/	/
Malt	(*cf. Beer, hop and malt*)	(*cf. Beer, hop and malt*)
Necessary and extra necessary money	£11.921/12/4	£12,605/12/1
Oats	(*cf. Grotts, oatmeal, oates*)	(*cf. Grotts, oatmeal, oates*)
Oil	£53,376/13/11	/
Peases	£59,742/1/0	/
Pork	/	/
Promiscuous payments	£4,057/17/10	£148/1/10
Provisions	£198,343/11/10	£9,143/14/6
Rent	£811/17/0	£527/12/0
Salary	£2,976/17/7	£7,347/17/2
Salt	£4,953/18/2	/

7 Items modified from previous table have been italicised.
8 Register abstract in ADM 20/246.
9 Register abstract in ADM 22/248.

Stationery ware	/	/
Staves	(*cf.* *Casks, hoops and staves*)	(*cf.* *Casks, hoops and staves*)
Vinegar	£1,200/9/3	/
Wheat	£32,405/3/6	£649/10/5
Wine	£28,202/11/7	/
Workmanship	£4,571/0/11	£1,254/10/3
Extra payments	£225/0/0	£503/11/2
Wages	/	/
Total	**£524,294/13/2**	**£44,955/13/1**

APPENDIX 5

List of Victuals on the Southsea Castle *Leaving for the East Indies at the End of 1759 with a Crew of 130 Men*

- Bread, 8 months of the newest diet and best in store and what can be stow in the bread room to be packed in iron bound casks.
- Beer as much as it can conveniently stow.
- Beef plus a monthly of the newest in store including 4 months flour and suet.
- Pork 8 months of the newest, suet 4 months without flour.
- Peases 8 months of the newest kiln dried.
- Grotts 2 months.
- Stockfish 4 months in lieu of grotts at the rate of 2 fishs and 4 pounds weigh of a gallon of grotts.
- 4 months kiln dried peas in lieu of four months due proportion of grotts to be issued in lieu of oatmeal.
- Put into casks: – butter 2 months, cheese 2 months, oil 2 months.
- Sugar 3 months in lieu of oil at the rate at 1 pound of sugar to be a pint of oil or if no sugar to be had, that 3 months oil is to be issued and the sugar is to be of the best muscoved[1] sort.
- Vinegar 12 months.
- Mustard seed ground 1½ pound per man for the complement being 130.
- Truss hoops nailed two setts.
- Iron hoops spare 140.
- Rivetts double allowance for the said hoops and other uses.
- Coopers tools and stores at double allowance.
- Water casks as many as she can stow.
- Barricoes[2] 16 N.

Source: PRO, ADM 111/49, 26 December 1759.

1 Muscoved sugar = muscovado sugar.
2 Barricoes = barrico, a keg or small cask.

APPENDIX 6

Process to Be Used in Curing Beef and Pork

For Beef

The oxen are to be haltered in the slaughterhouse (and not suffered to run loose about the same, as hath been heretofore practised to the great prejudice of the meat), and to hang twenty-four hours after slaughtered to cool before cut out. So the four quarters being weighed, and having taken out the four marrow bones and four shin bones, for which eight pound is allowed in each [*cwt*], every hundredweight is to be cut into twenty-six four-pound pieces, the randers[1] and messers[2] taking care it be done with as little waste and scraps as may be: thus messed, to be well rubbed with a mixture of two-thirds white and one-third French Bay salt between the layers, where they are to lie five or six days, but rather five; and to be taken taken out on the tables and well rubbed again with two-thirds Bay and one-third white salt between each layer, every other layer being trod close down; which cask being bunged up and made very tight, are to remain so for at least twelve days, after which they are to be laid with their bung downwards twenty-four hours to drain. And so the pickle is is to be put to them, that being well boiled and clean scummed, and so strong that a four-pound piece of beef will swim on the top when cold. The letter B for beef, with the number of each cask, and the number of pieces of beef in each cask, with the first two letters of the place where and month in which salted up, are to be put on the head of each cask.

For Pork

The hogs to hang twenty-four hours before carried to be weighed; to weigh three hogs at a time, and to cut fifty-five two-pound pieces to each hundredweight; the randers and messers to take care that they be cut clean from scraps, the pieces to be rubbed with like salt of each proportions and to continue in the bins and in the cask and draining the same time; and in like manner as directed for the beef, also to be numbered and marked respectively according to the direction therein, and you and your clerk is also to be present with the

1 Men who cut meat into strips (rands).

2 Men who cut meat into regular common sizes for the seamen's mess.

[victualling clerk of the] Cheque at the weighing, cutting and messing of all the beef and pork, and to keep a particular account thereof.

You are further to observe that no beef provided at this office be sent on board any of his Majesty's ships in pieces of any other weight that of four pounds (unless we shall order the cutting beef into three-pound pieces to be issued in lieu of pork), and that no pork be issued but in pieces of two-pounds each, and that at no time there be any unusual pieces put up with the other flesh, such as leg bones, shins of oxen, or the cheeks of hogs or ox hearts, etc. And to take care that the beef and pork cut in the cuttinghouse of this office do always hold such weight as that every twenty-eight pieces of beef cut for four-pound pieces, took out of the cask as they rise and the salt shaken off it, shall weigh one hundred pounds net avoirdupois weight, and every fifty-six pieces of salted pork, cut for two-pound pieces, took out of the cask and shaken as in the beef, shall weigh one hundred and four pounds net avoirdupois ...

Source: Baugh, *British Naval Administration*, 413–14, Instructions of 1716, confirmed in 1736, 1764, 1766 and 1784.

APPENDIX 7

Wage Totals, According to Activity, Paid to Victualling Personnel in the London Yard in the First Quarter of 1761

Coopers		£1,779/4/10
Frying joynters		£15/11/3
Millers at Rotherhithe		£297/14/1
Bakers at Rotherhithe		£430/8/7
Bag menders		£33/11/10
Millers at Redhouse		£131/19/10
Brewers at Redhouse		£507/5/10
Labourers at Redhouse		£912/8/8
Gate keepers and messengers		£177/7/6
Cooperage labourers		£574/4/0
Pickle yard labourers		£435/1/8
Bakehouse labourers		£38/3/6
Wharf labourers		£212/7/2
Cartage by wheeler/tickett		£299/13/4
Waterage and petty freight		£478/19/0
Sweeping chimney and destroy vermin		£5/12/0
Messers and salter		£923/1/8
Princess Augusta (7 men)		£35/14/3
Caesar Sloop (5 men)		£28/18/6
	Total	**£8,467/12/2**

Source: PRO, ADM 113/144.

APPENDIX 8

Supervisory Staff of the Victualling Board, 1761

March 2nd 1761

To John Clevland Esq. [Secretary to the Admiralty Board]

In return to your letter.

List of the officers and how long they have served and board opinion as to their qualification.

Officer's name	Employment	Age	Date of appointment
London			
Denham Skeet	Accountant for Cash	57	10 April 1738
Denham Brigg	Chief Clerk for examining and stating Imprest Accounts	36	22 February 1757
Robert Hall	Chief Clerk for keeping a charge on the Treasurer and to assign Bills	41	9 June 1749
Richard Henshaw	Accountant for Stores	44	31 October 1752
Thomas Dyals	Chief Clerk for keeping a charge on Pursers and adjusting accounts with Masters of Transports	40	31 October 1752
Gaston Martineau	Clerk of Issue	65	31 October 1752
Edward Tymewell	Chief Clerk under the Accountant for Stores for stating, passing and balancing Pursers' Accounts		
James Butcher	Master Cooper	44	23 December 1746
James Wood	Clerk of the Cutting House	60	27 August 1736
Thomas Hayward	Clerk of the Bakehouse and Manager of the Mills at Rotherhithe	53	15 December 1747
Joshua Deardon	Clerk of the Brewhouse	50	5 April 1756
John Raymond	Master Brewer	48	8 December 1757
John Kelsall	Victualling Officer: Redhouse and Deptford	64	13 March 1756
Benjamin Granvill	Employed in surveying and certifying to the performance of the several artificers work	57	30 July 1730
John Harper	Muster Master of the labourers etc.	64	30 March 1725
Alexander St-Barbe	Hoytaker	45	8 January 1757
Johne Everall	Necessary and porter	53	30 November 1744
Portsmouth			
Matthew Oaks	Agent	41	14 February 1757
George Bray	Storekeeper	58	5 July 1724
Leonard Lloyd	Clerk of the Cheque	47	23 February
William Rogers	Master Cooper	65	30 December 1744
William Philips	Master Brewer	43	9 October 1751

Plymouth			
John Ommanney	Agent	58	12 December 1754
John Watkins	Storekeeper	30	12 March 1760
Alex Gordon	Clerk of the Cheque	53	14 August 1759
Richard Pett	Master Cooper	67	23 October 1751
Edward Williams	Master Brewer	43	1 October 1756

Chatham			
Henry Hutchinson	Agent	60	27 February 1756
Robert Russel	Storekeeper	65	18 September 1759

Dover			
Michael Russel	Agent	49	15 December 1752
I. Hammond	Storekeeper	46	8 August 1748
Stephen Semphire	Clerk of the Cheque	46	7 September 1748
Benjamin Worthington	Master Cooper	42	8 July 1753

Gibraltar			
John Hewes	Agent		1 August 1747

Source: PRO, ADM 110/20, 2 March 1761.

Sources and Bibliography

Sources

The National Archives: Public Record Office

ADM 2: Orders and Instructions (out-letters) sent by the Admiralty.
(In alphabetical order according to recipient.)
ADM 2/76 (3 December 1755 – 18 June 1756)
77 (21 June 1756 – 27 January 1757)
78 (28 January – 23 June 1757)
79 (24 June – 31 December 1757)
80 (January – July 1758)
81 (August – December 1758)
82 (1 January – 11 May 1759)
83 (12 May – 31 October 1759)
84 (November 1759 – May 1760)
85 (June 1760 – 17 January 1761)
86 (19 January – 17 August 1761)
87 (18 August 1761 – 18 March 1762)
88 (9 March – 23 August 1762)
89 (24 August 1762 – 10 February 1763)
90 (21 February – 23 August 1763)

ADM 20: Treasurer's ledgers.
(These ledgers list all the Victualling Board's receipts and expenses in the given period)
ADM 20/225 (1756)
228 (1757)
231 (1758)
235 (1759)
238 (1766)
242 (1761)
244 (January–May 1762)
246 (June_December 1762)
248 (1763–64)

ADM 30: Miscellaneous.
(Ledger no. 44 is particularly interesting as it lists the various rules and regulations relating to curing procedures, storage procedures and boarding procedures)

ADM 109: Correspondence received by the Victualling Board (in-letters).
(This series did not begin until 1816. However, correspondence exchanged with the Lords of the Admiralty can be found in the ADM C series which is housed in the National Maritime Museum in Greenwich)

ADM 110: Correspondence emanating from the Victualling Board (out-letters)
ADM 110/18 (1755–57)
19 (1757–59)
20 (1760–62)
21 (1762–64

ADM 111: Minutes of the Victualling Board.
(This series lists the names of those merchants who participated in the invitations to tender bids and the tariffs that were concluded. However, the minutes of the last quarter of 1760 are missing)
ADM 111/43 (October 1755 – March 1756)
44 (April – September 1756)
45 (October 1756 – May 1757)
46 (May – December 1757)
47 (January – August 1758)
48 (August 1758 – March 1759)
49 (April 1759 – December 1760)
50 (1760)
51 (January – June 1761)
52 (July – December 1761)
53 (January – June 1762)
54 (July – December 1762)

ADM 112: Correspondence from Victualling Board agents to the Commissioners.
(A very useful series for understanding the contracts in detail. However, there are only a very few documents in it relating to the War of the Austrian Succession, and no documents relating to the Seven Years War. The series only becomes detailed from 1776)

ADM 113: Victualling yards pay-list (quarterly breakdown).
ADM 113/12, 13, 14 Chatham (1752–64)
35 Cork (1758–60)
58 Deptford (1756–61)
89, 90, 91 Dover (1752–69
108, 109 Gibraltar (1747–63)
124 East Indies (1758–63)
142, 143, 144, 145 London (1756–64)
185, 186, 187, 188 Plymouth (1755–63)
223, 224, 225, 226 Portsmouth (1752–69)
248 Rotherhithe (1762–93)

ADM 114: Victualling Board – Miscellaneous.
ADM 114/14 (1701–65)
35 (1723–65)

Corporation of London Records Office

Three distinct sources provide supplementary information on some suppliers:

1. Indexes of Admissions to the Freedom of the City

These ledgers list the names of the merchants, in alphabetical order, who were registered members of the Corporation of London Merchants. Beside each name appears the date on which the certificate of membership was delivered. They refer back to the CF1 archival series, in which the duplicates were kept. The certificates are of particular interest because they reveal certain personal details concerning these merchants, notably the place and profession of their fathers.

The names of the merchants only appear once in the index – namely on the date on which they joined the corporation – yet nineteen especially important contractors were successfully traced back from 1756 to 1701. They were R. Barnevelt, T. Bidwell, J. Biggin, J. Cook, R. Cowley, T. Farrar, E. Fossett, J. Le Keux, T. Lewis, S. Mellish, W. Millward, A Newman, C. Owen, W. Parker, J. Pasco, W. Preddy, E. Shallet, A. Spencer, T. Townsend.

The following references were successfully followed through:

CFI/473 Thomas Farrar (May 1727)
CFI/481 William Parker (June 1728)
CFI/507 William Millward (August 1730)
CFI/539 Richard Cowley (April 1733)
CFI/542 William Parker (July 1733)
CFI/743 William Preddy (June 1750)

2. A file concerning Abraham Newman was also found (MS 8621). It contains a number of letters that offer information regarding the merchant's business accounts in the 1766–99 period.

3. The Trade Directory lists the names and addresses of merchants in the London and Greater London area. Some supplementary information was obtained through the 1755, 1758 and 1760 directories.

Bibliography

Printed Primary Sources

Anon., *The Case of the Pursers of Her Majesties Navy* (London, 1711).
Baugh, D.A., ed., *Naval Administration 1715–1750* (London, Navy Records Society, vol. 120, 1977).
Bromley, J.S., *The Manning of the Royal Navy: Selected Public Pamphlets, 1693–1873* (London, Navy Records Society, vol. 119, 1974).

Calendar of State Papers Colonial, America and West Indies, Vol. 13 (*1689–92*), Vol. 14 (*1693–1696*), ed. J.W. Fortescue (London, H.M. Stationery Office, 1901, 1903); Vol. 21, *1702–1703*, Vol. 24, *1708–1709*, Vol. 26, *1711–12*, ed. C. Headlam (London, H.M. Stationery Office, 1913, 1922, 1925).

Calendar of State Papers Domestic, William and Mary, Nov. 1691–Dec. 1692, ed. W.J. Hardy (London, H.M. Stationery Office, 1900).

Cockburn, W., *An Account of the nature, causes, symptoms, and cure of the distempers that are incident to sea-faring people* (London, H. Newman, 1696), third edition published as *Sea Diseases* (London, G. Strahan, 1736).

Gwyn, J.R.J., ed., *The Royal Navy and North America: The Warren Papers, 1736–1752* (London, Navy Records Society, vol. 118, 1973).

Hodges, H.W. and E.A. Hughes, eds, *Select Naval Documents* (Cambridge, Cambridge University Press, 1922), 125–6.

Kemp, P., ed., 'Boscawen's Letter to His Wife, 1755–1756', in *Naval Miscellany*, IV, ed. C. Lloyd (London, Navy Records Society, vol. 92, 1953), 163–256.

Lind, J., *A Treatise of Scurvy* (Edinburgh, 1753).

Lind, J., *An Essay on the most effectual means of preserving the Health of Seamen* (London, new ed. 1774).

Lloyd, C.C., ed., *The Health of Seamen: Selections from the Works of Dr. James Lind, Sir Gilbert Blane and Dr. Thomas Trotter* (London, Navy Records Society, vol. 107, 1965).

Lloyd, C.C., *The British Seaman, 1200–1860: A Social History* (London: Collins, 1968).

MacIntyre, J., *A Military Treatise on the discipline of the Marine Forces* (London, 1763).

Mackay, R.F., ed., *The Hawke Papers: A Selection, 1743–1771* (Aldershot, Navy Records Society, vol. 129, 1990).

Matcham, M.E., *A Forgotten John Russel being letters to a man of business, 1724–1751* (London, Arnold, 1905).

Mead, R., *A Discourse on the Scurvy* (1749), published in *The Medical Works of Richard Mead M.D.* (Edinburgh, 1763).

Merriman, R.D., ed., *Queen Anne's Navy* (London, Navy Records Society, vol. 103, 1961).

Molyneux, T.C., *Conjunct Expeditions: or expeditions that have been carried out jointly by the army and navy* (London, 1759).

Ranft, B.M., *The Vernon Papers* (London, Navy Records Society, vol. 99, 1958).

Syrett, D., *The Siege and Capture of Havana 1762* (London, Navy Records Society, vol. 114, 1970).

Tanner, J.R., ed., *Two Discourses of the Navy 1638 and 1659 by J. Hollond* (London, Navy Records Society, vol. 7, 1896).

Tanner, J.R., ed., *A Descriptive Catalogue of the Naval Manuscripts in the Pepysian Library at Magdalene College*, I (London, Navy Records Society, vol. 26, 1903).

Thompson, W., *The Royal Navy-Men's Advocate: or the corrupt practise of victualling the Navy* (London, 1757).

Thompson, W., *An Appeal to the Public ... to prevent the Navy of England being supplies with Pernicious Provisions* (London, 1761).

Secondary Sources

Acerra, M., *Rochefort, La construction navale française (1661–1815)* (Paris, Librairie de l'Inde Editeur, 1993).

Acerra, M. and J. Meyer, *La Grande Epoque de la Marine à Voiles* (Rennes, Ouest-France, 1987).

Acerra, M., J. Merino and J. Meyer, eds, *Les Marines de Guerre Européennes, XVIIe–XVIIIe siècles* (Paris, Presses de l'Université de Paris-Sorbonne, 1985).

Allen, R., 'Agriculture during the Industrial Revolution', in *The Economic History of Great Britain since 1700*, ed. R. Floud and J. McCloskey (Cambridge, Cambridge University Press, 2nd edn, 1995), 96–122.

Appleby, A.B., *Famine in Tudor and Stuart England* (Liverpool, Liverpool University Press, 1959).

Appleby, A.B., 'Grain crises and subsistence crises in England and France, 1590–1740', *Journal of Economic History*, vol. XXXIX, no. 4 (1979), 865–87.

Ashton, T.S., 'Changes in standards of comfort in eighteenth century England', *Proceedings of the British Academy*, vol. XLI (1955), 171–87.

Ashton, T.S., *An Economic History of England. The 18th Century* (London, Methuen, 1955).

Ashton, T.S., *Economic Fluctuations in England, 1700–1800* (Oxford, Clarendon Press, 1959).

Baker, N., *Government and Contractors: The British Treasury and War Supplies 1775–1783* (London, Athlone Press, 1971).

Baugh, D.A., *British Naval Administration in the Age of Walpole* (Princeton, Princeton University Press, 1965).

Baugh, D.A., 'The eighteenth century Navy as a national institution', in *The Oxford Illustrated History of the Royal Navy*, ed. R. Hill (Oxford, Oxford University Press, 1995), 120–60.

Baugh, D.A., *The Global Seven Years War 1754–1763* (Harlow, Longman, 2011).

Beatson, R., *Naval and Military Memoirs of Great Britain from 1727 to 1783* (London, J. Strachan, 1804).

Bédarida, F., F. Crouzet, D. Johnson, *De Guillaume le Conquérant au Marché Commun. Dix siècles d'histoire franco-britannique* (Paris, A. Michel, 1979).

Bély, L., Y.M. Bercé, J. Bérenger, A. Corvisier, J. Meyer and R. Quatrefages, *Guerre et Paix dans l'Europe du XVIIe*, 3 vols (Paris, SEDES, 1991).

Bérenger, J., 'Les Habsbourg et la Mer', in *Mélanges offerts à Monsieur le Professeur Jean Meyer: Etat, Marine et Société*, ed. M. Acerra, J.-P. Poussou, M. Vergé-Franceschi and A. Zysberg (Paris, Presses de l'Université de Paris-Sorbonne, 1995), 25–34.

Berrill, K., 'International trade and the rate of economic growth', *Economic History Review*, new series, XII, no. 3 (1960–1), 351–9.

Beveridge, Sir W.H., *Prices and Wages in England from the Twelfth to the Nineteenth Century* (London, Frank Cass, 1939, repr. 1965).

Black, J. and P. Woodfine, *The British Navy and the Use of Naval Power in the Eighteenth Century* (Leicester, Leicester University Press, 1988).

Bosher, J.F., 'Financing the French Navy in the Seven Years' War: Beaujon, Goossens et Compagnie in 1759', *Business History*, 28 (1986), 115–33.

Bosher, J.F., 'Guerre et activités de la marine marchande du Canada (1743–1763)', in *Mélanges offerts à Monsieur le Professeur Jean Meyer: Etat, Marine et Société*, ed. M. Acerra, J.-P. Poussou, M. Vergé-Franceschi and A. Zysberg (Paris, Presses de l'Université de Paris-Sorbonne, 1995), 49–71.

Braudel, F., *Civilisation matérielle, Economie et Capitalisme, XVe–XVIIIe siècles*, vol. 3, *Le Tempes du Monde* (Paris, A. Colin, 1979).

Brewer, J., *The Sinews of Power: War, Money and the English State, 1688–1783* (London, Unwin Hyman, 1989).

Brown, H.P. and S.V. Hopkins, *A Perspective of Wages and Prices* (London, Methuen, 1981).

Bruchey, S., *The Colonial Merchants: Sources and Readings* (New York, Harcourt, 1966).

Buchet, C., 'Les levées d'hommes opérées par la Royal Navy aux Antilles entre 1689 et 1730', *Histoire, Economie et Société*, vol.10, no. 4 (1990), 521–43.

Buchet, C., *La lutte pour l'Espace Caraïbe et la Façade Atlantique de l'Amérique Centrale et du Sud, 1672–1763*, 2 vols (Paris, Librairie de l'Inde Editeur, 1991).

Buchet, C., 'Les modalités évolutives de la logistique anglaise en matériel naval dans l'espace caraïbe (1689–1783)', *Histoire, Economie et Société*, vol. 11, no. 4 (1992), 571–96.

Buchet, C., *Le Tour du Monde de l'Amiral Anson* (Paris, France-Empire, 1992).

Buchet, C., 'La logistique française en matériel naval dans l'espace caraïbe (1672–1763)', Comité de Documentation Historique de la Marine, Communication de 14 Janvier 1992 (Vincennes, Service Historique de la Marine, 1993), 43–74.

Buchet, C., 'Des routes maritimes Europe-Antilles et de leurs incidences sur la rivalité franco-britannique', *Histoire, Economie et Société*, vol. 13, no. 4 (1994), 563–82.

Buchet, C., 'L'Ordnance Board et la logistique en armes pour les flottes anglaises aux Antilles (1689–1763)', in *Mélanges offerts à Monsieur le Professeur Jean Meyer: Etat, Marine et Société*, ed. M. Acerra, J.-P. Poussou, M. Vergé-Franceschi and A. Zysberg (Paris, Presses de l'Université de Paris-Sorbonne, 1995), 73–87.

Buchet, C., 'Révélateurs d'une détermination coloniale: les arsenaux anglais de l'espace caraïbe', *Histoire, Economie et Société*, vol. 16, no. 1 (1997), 123–49.

Buchet, C., 'The development of Victualling Board bases in London, Portsmouth, Plymouth, Chatham and Dover (1701–1763)', *Transactions of the Naval Dockyards Society*, vol. 4 (2008), 53–68.

Butel, P., *Les Négociants bordelais, L'Europe et les Iles au XVIIIe siècle* (Paris, Aubier, 1974).

Butel, P., 'L'Apogée du Grand Commerce Maritime', in *Histoire Economique et Sociale du Monde*, ed. P. Léon, vol. 3 (Paris, A. Colin, 1978), 51–136.

Butel, P., *L'Economie Française au XVIIIe siècle* (Paris, SEDES, 1993).

Butel, P., 'La dynamique des marchés coloniaux au XVIIIe siècle. L'exemple britannique', in *Mélanges offerts à Monsieur le Professeur Jean Meyer: Etat, Marine et Société*, ed. M. Acerra, J.-P. Poussou, M. Vergé-Franceschi and A. Zysberg (Paris, Presses de l'Université de Paris-Sorbonne, 1995), 99–108.

Butel, P., 'L'Espace Atlantique au XVIIIe', *Revue Historique des Armées* (1996), 57–64.

Butel, P., *Histoire de l'Atlantique de l'Antiquité à nos jours* (Paris, Perrin, 1997).

Butel, P. and B. Lavallé, eds, *L'Espace Caraïbe, Théâtre et Enjeu des Luttes Impériales (XVIe–XIXe siècles)* (Bordeaux, Maison des Pays Ibériques, 1996).

Carpenter, K.J., *The History of Scurvy and Vitamin C* (Cambridge, Cambridge University Press, 1986).

Chalkin, C., *The Rise of the English Town 1750–1850* (Cambridge, Cambridge University Press, 2001).

Chambers, J.D. and G.E. Mingay, *The Agricultural Revolution in England 1750–1880* (London, Batsford, 1956).

Chaunu, H. and P., *Séville et l'Atlantique 1504–1650* (Paris, SEVPEN, 1955–60).

Chaunu, P., *La Civilisation de l'Europe des Lumières* (Paris, Arthaud, 1971).

Clapham, J.H., *The Bank of England*, vol. 1: *1694–1797* (Cambridge, Cambridge University Press, 1994).

Clark, G.N., 'War trade and trade war', *Economic History Review*, vol. 1, no. 2 (1928), 262–80.

Clark, G.N., *The Wealth of England from 1496 to 1760* (London, Oxford University Press, 1946).

Coad, J.G., *Historic Architecture of Chatham Dockyard, 1700–1850* (London, National Maritime Museum and Society of Nautical Research, 1982).

Coad, J.G., *Historic Architecture of H.M. Naval Base Devonport, 1689–1850* (London, National Maritime Museum and Society of Nautical Research, 1983).

Coad, J.G., *The Royal Dockyards 1690–1850* (Aldershot, Scolar Press, 1989).

Coleman, D.C., 'Naval dockyards under the later Stuarts', *Economic History Review*, 2nd series, 6 (1953), 134–55.

Coleman, D.C., *The Economy of England, 1450–1750* (London, Oxford University Press, 1977).

Corbett, J.S., *England in the Seven Years' War: A Study in Combined Strategy* (London, Longmans, 1907).

Corvisier, A., 'La Place de l'Armée et de la Marine dans les Institutions, l'Economie et la Société', *Bulletin de la Société d'Histoire Moderne et Contemporaine*, 12 (1980), 141–55.

Crafts, N.F.R., 'British economic growth 1700–1831: a review of the evidence', *Economic History Review*, XXXVI, no. 2 (May 1983), 177–99.

Crewe, D.G., *Yellow Jack and the Worm. British Naval Administration in the West Indies, 1739–1748* (Liverpool, Liverpool University Press, 1993).

Crouzet, F., 'L'Angleterre et la France au XVIIIe siècle: essai d'analyse comparée de deux croissances économiques', *Annales E.S.C.*, vol. 21 (1966), 254–91.

Crouzet, F., *De la supériorité de l'Angleterre sur la France, l'Economique et l'Imaginaire XVIIème–XXème siècles* (Paris, Perrin, 1985).

Crouzet, F., 'Remarques sur la formation d'une économie mondiale', *Histoire, Economie et Société*, 4 (1986), 611–25.

Cullen, L.M., *An Economic History of Ireland since 1660* (London, Batsford, 1972).

Cullen, L.M. and P. Butel, eds, *Négoce et industrie en France et en Irlande aux XVIIIe et XIXe siècles* (Bordeaux, CNRS, 1980).

Davies, M.G., 'The country gentry and payments to London, 1650–1714', *Economic History Review*, XXIV (1971), 15–36.

Davis, R., 'English foreign trade, 1770–1774', *Economic History Review*, XV (1962), 285–303.

Davis, R., *The Rise of the English Shipping Industry in the XVIIth and XVIIIth centuries* (2nd edn, Newton Abbot, David and Charles, 1972).

Deane, P. and W.A. Cole, *British Economic Growth, 1688–1959: Trends and Structure* (Cambridge, Cambridge University Press, 1962).

Dickson, D., 'The Cork merchant community in the eighteenth century', in *Négoce et Industrie en France et en Irlande*, ed. L.M. Cullen and P. Butel (Bordeaux, CNRS, 1980), 45–50.

Dickson, P.G.M., *The Financial Revolution in England: A study in the development of public credit 1688–1756* (London, Macmillan, 1967).

Drummond, J.C. and A. Wilbraham, *The Englishman's Food. Five centuries of English diet* (London, Jonathan Cape, 1939).

Duffy, M., 'The creation of Plymouth Dockyard and its impact on naval strategy', in *Guerres Maritimes (1688–1713)* (Vincennes, Service Historique de la Marine, 1996), 245–74.

Ehrman, J., *The Navy in the War of William III, 1689–1697: Its State and Direction* (Cambridge, Cambridge University Press, 1953).

Engerman, S.L., 'Mercantilism and overseas trade', in *The Economic History of Britain since 1700*, ed. R. Floud and D. McCloskey, vol. I (2nd edn, Cambridge, Cambridge University Press, 1995), 182–204.

Eversley, D.E.C., 'The home market and economic growth in England 1750–1780', in *Land, Labour and Population in the Industrial Revolution*, ed. J.D. Chambers and G.E. Mingay (London, Edward Arnold, 1967), 206–59.

Falconer, W., *An Universal Dictionary of the Marine* (London, 1771).

Farnie, D.A., 'The commercial empire of the Atlantic 1607–1783', *Economic History Review*, 2nd series, vol. XV, no. 2 (1962), 205–18.

Fisher, F.J., *London and the English Economy, 1500–1700*, ed. P.J. Corfield and N.B. Harte (London, Hambledon Press, 1990).

Fussell, G.E., 'The change in farm labourers' diet during two centuries', *Economic History*, 1 (1927), 269–74.

Fussell, G.E., 'The London cheesemongers of the eighteenth century', *Economic History*, 3 (Jan. 1928), 394–8.

Fussell, G.E. and C. Goodman, 'Eighteenth-century traffic in live-stock', *Economic History*, XI (Feb. 1936), 214–36.

Fussell, G.E. and C. Goodman, 'The eighteenth century traffic in milk products', *Economic History*, XII (Feb. 1937), 380–7.

Fussell, G.E. and C. Goodman, 'Traffic in farm products in eighteenth-century England', *Agricultural History*, XII (1938), 355–68.

George, M.D., *London Life in the Eighteenth Century* (London, Paul Kegan, 1925).

Glete, J., *Navies and Nations: Warships, Navies, and State Building in Europe and America, 1500–1860*, 2 vols (Stockholm, Almqvist, 1993).

Gradish, S.F., *The Manning of the British Navy during the Seven Years War* (London, RHS, 1980).

Hamilton, R.V., *Naval Administration: The Constitution, Character, and Functions of the Board of Admiralty and the Civil Departments It Directs* (London, Royal Navy Handbooks, Bell, 1896).

Hancock, D., *Citizens of the World. London Merchants and the Integration of the British Atlantic Community (1735–1785)* (Cambridge, Cambridge University Press, 1995).

Hattendorf, J.B., *England in the War of Spanish Succession: A Study of the English View and Conduct of Grand Strategy, 1702–1712* (New York, Garland, 1987).

Hattendorf, J.B., 'The struggle with France, 1690–1815', in *The Oxford Illustrated History of the Royal Navy*, ed. R. Hill (Oxford, Oxford University Press, 1995), 80–119.

Haudrère, P., *La Compagnie Française des Indes au XVIIIe siècle (1719–1795)* (Paris, Librairie de l'Inde Editeur, 1989).

Haudrère, P., 'L'Etat et les négociants: deux memoires sur la situation de la Compagnie Française des Indes en 1730', in *Mélanges offerts à Monsieur le Professeur Jean Meyer: Etat, Marine et Société*, ed. M. Acerra, J.-P. Poussou, M. Vergé-Franceschi and A. Zysberg (Paris, Presses de l'Université de Paris-Sorbonne, 1995), 209–19.

Haudrère, P., *L'Empire des Rois 1500–1789* (Paris, Denoël, 1997).

Holderness, B.A., *Pre-Industrial England. Economy and Society from 1500–1750* (London, Dent, 1976).

Holmes, G., *The Making of a Great Power: Pre-Industrial Britain 1660–1783* (London, Longman, 1982).

Howard, M.E., *The British Way in Warfare: A Reappraisal* (London, Cape, 1975).

Jackson, G., *Hull in the XVIIIth Century* (London, Oxford University Press, 1972).

Jenkins, R., 'Copper smelting in England', *Transactions of the Newcomen Society*, vol. XXIV (1943–5), 73–80.

John, A.H., 'War and the English economy, 1700–1763', *Economic History Review*, new series, 7 (1954–5), 329–44.

John, A.H., 'The London company and the marine insurance market of the eighteenth century', *Economica*, XXV (1958), 126–41.

John, A.H., 'Aspects of English economic growth in the first half of the eighteenth century', *Economica*, XXVIII (1961), 176–90.

Jones, E.L., *Seasons and Prices. The role of the weather in English agricultural history* (London, Allen & Unwin, 1964).

Jones, E.L., *Agriculture and Economic Growth in England, 1650–1815* (London, Methuen, 1967).

Jones, E.L., 'English and European agricultural development 1650–1750', in *The Causes of the Industrial Revolution in England*, ed. R.M. Hartwell (London, Methuen, 1967).

Kennedy, P.M., *The Rise and Fall of British Naval Mastery* (London, Allen Lane, 1976).

Kennedy, P.M., *The Rise and Fall of the Great Powers: Economic Change and Military Conflict from 1500 to 2000* (London, Unwin, 1988).

Lavery, B., *The Ship of the Line*, 2 vols. I: *The Development of the Battle Fleet, 1650–1850*. II: *Design, Construction, and Fittings* (London, Conway, 1983–4).

Legohérel, H., *Les Trèsoriers Généraux de la Marine (1715–1788)* (Paris, Editions Cujas, 1965).

Léon, P., ed., *Histoire Economique et Sociale du Monde*, vol. II, *Les hesitations de la croissance 1580–1740*, vol. III, *Inerties et révolutions* (Paris, Armand Colin, 1978).

Lloyd, C., 'Victualling of the fleet in the 18th and 19th centuries', in *Starving Sailors*, ed. J. Watt, E.J. Freeman and W.F. Bynum (London, National Maritime Museum, 1981).

Lloyd, C. and J.L.S. Coulter, *Medicine and the Navy*, vol. III (Edinburgh, Livingstone, 1961).

McCusker, J. and J. Menard, *The Economy of British America 1607–1789* (Chapel Hill, University of North Carolina Press, 1985).

Macdonald, J., 'A new myth of naval history? Confusing magnitude with significance in British naval victualling purchases, 1750–1815', *International Journal of Maritime History*, XXI (2009), 159–88.

Mackay, R.F., *Admiral Hawke* (Oxford, Clarendon Press, 1965).

McKay, D. and H.M. Scott, *The Rise of the Great Powers 1648–1815* (London, Longman, 1983).

Mahan, A.T., *The Influence of Sea Power upon History* (London, Sagamore Press, 1957).

Maitland, W., *The History and Survey of London* (London, 1756).

Mante, T., *The History of the War in North America and the Islands of the West Indies* (London, W. Strahan, 1772).

Marcus, G.J., *Heart of Oak: A Survey of British Sea Power in the Georgian Era* (London, Oxford University Press, 1975).

Mathias, P., *The Brewing Industry in England 1700–1830* (Cambridge, Cambridge University Press, 1959).

Mathias, P., *The First Industrial Nation. An economic history of Britain, 1700–1914* (2nd edn, London, Methuen, 1983).

Mathias, P. and P. O'Brien, 'Taxation in Britain and France, 1715–1810. A comparison of the social and economic incidence of taxes collected for the central governments', *Journal of European Economic History*, 5 (1976), 601–50.

Meyer, J., *L'Armement nantais dans la deuxième moitié du XVIII*[e] *siècle* (Paris, SEVPEN, 1969).

Meyer, J., *L'Europe et la Conquête du Monde (XVI–XVIIe siècles)* (New York and Paris, A. Colin, 1975).

Meyer, J., *Les Capitalismes* (Paris, PUF, 1980).

Meyer, J., *Le Poids de l'Etat* (Paris, PUF, 1983).

Middleton, R., *The Bells of Victory: The Pitt-Newcastle Ministry and the Conduct of the Seven Years' War, 1757–1762* (Cambridge, Cambridge University Press, 1985).

Minchinton, W.E., *The Growth of English Overseas Trade in the Seventeenth and Eighteenth Centuries* (London, Methuen, 1969).

Morgan, K., *Bristol and the Atlantic Trade in the Eighteenth Century* (Cambridge, Cambridge University Press, 1993).

Morineau, M., *Les Faux-semblants d'un demarrage économique, agriculture et demographie* (Paris, A. Colin, 1971).

Morineau, M., *Pour une histoire économique vraie* (Lille, Presses de l'Université de Lille, 1985).

Mousnier, R., 'L'évolution des finances publiques en France et en Angleterre pendant les guerres de la Ligue d'Augsburg et la Succession d'Espagne', *Revue Historique*, 44 (1951), 1–23.

Mullett, F.C., 'The cattle distemper in mid-eighteenth century England', *Agricultural History*, XX (1946), 144–65.

Namier, L. and J. Brooke, eds, *History of Parliament: The House of Commons 1754–1790*, 3 vols (London, History of Parliament Trust, 1964).

Neal, L.D., 'Interpreting power and profit in economic history: a case study of the Seven Years' War', *Journal of Economic History*, XXXVII, no. I (March 1977), 20–35.

Neal, L.D., 'The cost of impressment during the Seven Years' War', *Mariners Mirror*, 64 (February 1978), 45–56.

O'Brien, P.K., 'La croissance économique (XVIIIe–XXe siècles)', in *De Guillaume le Conquérant au Marché Commun. Dix siècles d'histoire franco-britannique*, ed. F. Bédarida, F. Crouzet and D. Johnson (Paris, 1979), 193–207.

O'Brien, P.K., 'Agriculture and the home market for English industry, 1660–1820', *English Historical Review*, 100 (1985), 773–99.

O'Brien, P.K., 'Central government and the economy, 1688–1715', in *The Economic History of Britain since 1700*, ed. R. Floud and D. McCloskey, vol. I (2nd edn, Cambridge, 1995), 205–41.

O'Donovan, J., *The Economic History of the Live Stock in Ireland* (Dublin and Cork, Cork University Press, 1940).

Oppenheim, M., *A History of the Administration of the Royal Navy ... from MDIX to MDCLX* (London, J. Lane, 1896).

Palmer, S. and G. Williams, eds, *Charted and Uncharted Waters* (London, University of London, 1982).

Pares, R., *War and Trade in the West Indies 1739–1763* (Oxford, Clarendon Press, 1936).

Pares, R., *Colonial Blockade and Neutral Rights, 1739–1763* (Oxford, Clarendon Press, 1938).

Pares, R., *Yankees and Creoles* (London, Longman, Green & Co., 1956).

Parry, J.H., *Trade and Dominion. The European Overseas Empire in the Eighteenth Century* (London, Wiedenfeld & Nicolson, 1971).

Penson, L.M., 'The London West India interest in the eighteenth century', *English Historical Review*, 36 (1921), 373–92.

Pool, B., *Navy Board Contracts 1660–1832* (London, Longman, 1966).

Porter, R., *English Society in the Eighteenth Century* (London, Allen Lane, 1982).

Poussou, J-P., *Bordeaux et le Sud-Ouest au XVIIIe. Croissance économique et attraction urbaine* (Paris, Ecole des Hautes Etudes en Sciences Sociales, J. Touzot, 1983).

Poussou, J-P., 'Les villes anglaises, du milieu du XVIIIème siècle à la fin du XVIIIème siècle', in J-P. Poussou, ed., *Les Villes en Europe Occidentale* (Paris, CDU, 1983), vol. II, 1–212.

Poussou, J-P., *Les Iles Britanniques, les Provinces-Unies, la Guerre et la Paix au XVIIe siècle* (Paris, Economica, 1991).

Poussou, J-P., 'Avertissement, préface et note', G.M. Trevelyan, *Histoire Sociale de l'Angleterre* (Paris, Robert Laffont, 1993).

Poussou, J-P., *La Terre et les Paysans en France et en Grande-Bretagne aux XVIIe et XVIIIe siècles* (Paris, SEDES, 1999).

Price, J.M., 'Economic function and growth of American port towns in the eighteenth century', *Perspectives in American History*, VIII (1974), 152–63.

Pritchard, J.S., *Louis XV's Navy, 1748–1762: A Study of Organization and Administration* (Kingston, McGill-Queen's University Press, 1987).

Richmond, H.W., *The Navy in the War of 1739–1748*, 3 vols (Cambridge, Cambridge University Press, 1920).

Riley, J.C., *International Government Finance and the Amsterdam Capital Market* (Cambridge, Cambridge University Press, 1980).

Riley, J.C., *The Seven Years War and the Old Regime in France: The Economic and Financial Toll* (Princeton, Princeton University Press, 1986).

Rodger, N.A.M., *The Admiralty* (Lavenham, Suffolk, Terence Dalton, 1979).

Rodger, N.A.M., *The Wooden World* (London, Collins, 1986).

Rodger, N.A.M., *A Naval History of Britain*. I: *The Safeguard of the Sea, 660–1649* (London, HarperCollins, 1997), II: *Command of the Ocean, 1649–1815* (London, Allen Lane-Penguin, 2004).

Rodger, N.A.M., 'The victualling of the British Navy in the Seven Years' War', *Bulletin du Centre d'Histoire des Espaces Atlantiques*, new series, 2 (1985), 37–54 (reprinted in Rodger, *Essays in Naval History, from Medieval to Modern* (Farnham, Ashgate, 2009), VII).

Rodger, N.A.M., 'La mobilisation navale au XVIIIe siècle', in *Mélanges offerts à Monsieur le Professeur Jean Meyer: Etat, Marine et Société*, ed. M. Acerra, J.-P. Poussou, M. Vergé-Franceschi and A. Zysberg (Paris, Presses de l'Université de Paris-Sorbonne, 1995), 365–74 (English language version in Rodger, *Essays in Naval History, from Medieval to Modern* (Farnham, Ashgate, 2009), IX).

Rostow, W.W., *How It All Began. The origins of the modern economy* (New York, McGraw-Hill, 1975).

Rule, J., *The Vital Century: England's developing economy, 1714–1815* (London, Longman, 1992).

Rutter, O., *A Brief History of the Firm of Messrs. Davison, Newman and Company* (London, Davison, Newman & Co., 1938).

Ryan, A.N., 'The Royal Navy and the blockade of Brest, 1689–1805: theory and practice', in *Les Marines de Guerre Européennes (XVII–XVIIIe siècles)*, ed. M. Acerra, J. Merino and J. Meyer (Paris, PUPS, 1985), 175–93.

Schumpeter, E.B., 'English prices and public finance 1660–1822', *Review of Economic Statistics*, 20 (1938), 21–37.

Schumpeter, E.B., *English Overseas Trade Statistics, 1697–1808* (Oxford, Clarendon Press, 1960).

Schwarz, L.D., *London in the Age of Industrialisation, 1750–1850* (Cambridge, Cambridge University Press, 1992).

Scoville, W.C., *The Persecution of the Huguenots and French Economic Development, 1680–1720* (Berkeley, University of California Press, 1960).

Sharpe, J.A., *Early Modern England, a Social History, 1550–1760* (New York, Arnold, 1987).

Shepherd, J.F. and G.M. Walton, *Shipping, Maritime Trade and the Development of Colonial North America* (Cambridge, Cambridge University Press, 1972).

Sheridan, R.B., *Sugar and Slavery: An Economic History of the British West Indies 1623–1775* (Barbados, Caribbean University Press, 1974).

Skeel, C., 'The cattle trade between Wales and England from the fifteenth to the nineteenth centuries', *Transactions of the Royal Historical Society*, 4th series, IX (1926), 135–98.

Smith, A., *Enquête sur la Nature et les Causes de la Richesse des Nations* (Paris, PUF, 1995).

Sperling, J., 'The international payments mechanism in the seventeenth and eighteenth centuries', *Economic History Review*, XIV, no. 3 (1962), 446–8.

Syrett, D., 'American provincials and the Havana campaign of 1762', *New York History*, 49 (Oct. 1968), 375–90.

Syrett, D., 'The British landing at Havana: an example of an eighteenth century combined operation', *Mariner's Mirror*, 55 (August 1969), 325–31.

Syrett, D., 'The methodology of British amphibious operations during the Seven Years and American Wars', *Mariner's Mirror*, 58 (August 1972), 269–80.

Syrett, D., *Shipping and Military Power in the Seven Years War* (Exeter, Exeter University Press, 2008).

Tarrade, J., *Le Commerce Colonial de France à la fin de l'Ancien Régime, l'évolution du régime de l'Exclusif de 1763 à 1789* (Poitiers, Oudin et Beaulu, 1972).

Thirsk, J., *Economic Policy and Projects: The development of a consumer society in early modern England* (Oxford, Clarendon Press, 1978).

Thirsk, J., ed., *The Agrarian History of England and Wales 1500–1640*, vol. V, part 1 (Cambridge, Cambridge University Press, 1984).

Tracy, N., *Navies, Deterrence, and American Independence: Britain and seapower in the 1760s and 1770s* (Vancouver, British Columbia University Press, 1988).

Tunstall, B., *William Pitt, Earl of Chatham* (London, Hodder and Stoughton, 1938).

Turner, M., 'Agricultural productivity in England in the eighteenth century: evidence from crop yields', *Economic History Review*, XXXV (Nov. 1982), 489–510.

Van der Wee, H., 'Monetary credit and banking systems', in *The Cambridge Economic History of Europe*, ed. E.R. Ricj and C.H. Wilson, vol. 5 (Cambridge, Cambridge University Press, 1977), 290–392.

Vergé-Franceschi, M., *Les Officiers Généraux de la Marine Royale*, 15 vols (Paris, Librairie de l'Inde Editeur, 1990).

Vergé-Franceschi, M., ed., *Guerre et Commerce en Méditerranée (IX–XXe siècles)* (Paris, Kronos, 1991).

Vergé-Franceschi, M., *La Marine Française au XVIIIe siècle* (Paris, SEDES, 1996).

Vergé-Franceschi, M., 'Marine et opinion', *Revue Historique des Armées* (1996), 15–22.

Watt, J., E.J. Freeman and W.F. Bynum, eds, *Starving Sailors: The Influence of Nutrition upon Naval and Maritime History* (London, National Maritime Museum, 1981).

Weber, M., *L'Ethique Protestante et l'Esprit du Capitalisme* (Paris, Plon, 1985).

Westerfield, R.B., *The Middleman in English Business: particularly between 1660–1760* (New Haven, Yale University Press, 1915).

Willan, T.S., *The English Coasting Trade, 1660–1750* (Manchester, Publications of the University of Manchester, 1938).

Williams, G., *The Expansion of Europe in the Eighteenth Century* (London, Blandford Press, 1966).

Williams, J.E., 'Whitehaven in the eighteenth century', *Economic History Review*, VIII, no. 3 (1956), 393–404.

Williams, J.B., *British Commercial Policy and Trade Expansion, 1750–1830* (Oxford, Oxford University Press, 1976).

Wilson, C.H., *England's Apprenticeship 1603–1763* (London, Longmans, 1965).

Wilson, C.H., *Anglo-Dutch Commerce and Finance in the Eighteenth Century* (Cambridge, Cambridge University Press, 1966).

Wilson, C.H. and A. Carter, 'Dutch investment in eighteenth century England', *Economic History Review*, XII, no. 3 (April 1960), 434–9.

Wrigley, E.A., *Continuity, Chance and Change: The character of the Industrial Revolution in England* (Cambridge, Cambridge University Press, 1988).

Wrigley, E.A. and J. Hoppit, *The Industrial Revolution in Britain* (Oxford, Blackwell, 1994).

Wrigley, E.A. and R.S. Schofield, *The Population History of England: A reconstruction, 1541–1871* (Cambridge, E. Arnold, 1981).

Index

Note: page references to tables are in *italics*; those to figures are in **bold**

Other Works of Christian Buchet

In English

Ocean, (with Philip Plisson), Thames, 2006

In French

La lutte pour l'espace atlantique de l'Amérique Centrale et du Sud entre 1672 et 1763, Librairie de l'Inde, 1991 (Prize awarded by l'Académie des Sciences morales et politiques, l'Académie de Marine and l'Académie des Sciences d'outre-mer)

Le Tour du monde de l'amiral Anson, le "père de la Royal Navy", France-Empire, 1992

La Découverte de Tahiti, Preface by Paul-Emile Victor, France-Empire, 1993 (La Pérouse Prize of the Société de Géographie humaine de Paris)

Chef de guerre chez les indiens Creeks, France-Empire, 1994

L'Homme, la santé et la mer, Honoré Champion, 1999 (Prize awarded by l'Académie Nationale de medicine)

Marine, économie et société: un exemle d'interaction, l'avitaillement de la Royal Navy durant la guerre de Sept Ans, Honoré Champion, 1997 (Prize awarded by l'Académie de Marine)

Le Naufrage, (with Claude Thomasset), Honoré Champion, 1999

Sous la mer: le sixième continent, Presses de l'Université de Paris-Sorbonne, 2001

Marins et flibustiers du Roi-Soleil, (with Jean-Yves Nerzic, Pyré-Graph Editions, 2002 (Prix de l'Acoram 2003)

Les Voyous de la mer – naufrage, pollution, sécurité: le bilan santé de la mer, Ramsay, 2003

La Puissance maritime, (with Jean Meyer and Jean-Pierre Poussou), Presses de l'Université de Paris-Sorbonne, 2004

Une autre histoire des oceans at de l'homme, Preface by President Jacques Chirac, Robert Laffont, 2004 (Literary prize of the Cercle de la Mer 2004)

La Mer, la France et l'Amérique latine, Presses de l'Université de Paris-Sorbonne, 2006

La Mer, avenir de la Terre, photography by Philip Plisson, La Martinière, 2006, new edition 2009

Voyage autour du monde d'un père et d'un fils, Editions Origines, 2012

Printed and bound by CPI Group (UK) Ltd, Croydon, CR0 4YY

13/09/2026

14961003-0005